IT IS QUITE ANOTHER
ELECTRICITY

IT IS QUITE ANOTHER ELECTRICITY

TRANSMITTING BY ONE WIRE AND WITHOUT GROUNDING

SECOND EDITION, REVISED

MICHAEL BANK

PARTRIDGE

Print information available on the last page.

To order additional copies of this book, contact
Toll Free 800 101 2657 (Singapore)
Toll Free 1 800 81 7340 (Malaysia)
orders.singapore@partridgepublishing.com

www.partridgepublishing.com/singapore

CONTENTS

It is possible to transmit electric current over one wire, and, therefore, without necessary changes of the frequency and signal waveform, as well as without a use of the ground as the second wire.

It is possible to achieve zeroing (grounding) without inserting current into the ground. Perhaps grounding is achieved via an antenna.

The SWER System is an unbalanced system. As strange as it may sound, the proposed single-wire line is a balanced line.

A three-phase signal can be present at the input and output of a single-wire line.

It cannot be, because it can be never.
  —Anton Chekhov

CHAPTER 1

INTRODUCTION FROM A HISTORIC POINT OF VIEW

The War of Currents (sometimes called War of the Currents or Battle of Currents) was a series of events surrounding the introduction of competing electric power transmission systems in the late 1880s and early 1890s. This included commercial competition, a debate over electrical safety, and a media/propaganda campaign that grew out of it. The main players were the direct current (DC)-based Edison Electric Light Company and the alternating current (AC)-based Westinghouse Electric Company.

The method of AC received large support after the invention of the three-phase system in 1980.[1]

The three-phase system prevailed and has essentially been in use for the last 120 years following Dolivo-Dobrovolsky's invention.

Dolivo-Dobrovolsky

Dolivo-Dobrovolsky was a Russian engineer from my city, St. Petersburg.

---

1   See US Patent, publication number US427978 A, publication date May 13, 1890, by inventor Michael Von Dolivo-Dobrowolsky.

The first triumph of his three-phase system was displayed in Europe at the International Electro-Technical Exhibition of 1891. He used the system to transmit electric power at a distance of 176 kilometres with 75 per cent efficiency. This marked the beginning of an intensive implementation of the three-phase system. Still today, the majority of electrical energy is generated and distributed by the three-phase systems.

However, its capacity for an efficient generator and an efficient motor might be the only advantage of the three-phase system. While technology has changed over the last 120 years, the three-phase system has not changed significantly. It is now clear that this system has many problems and only one advantage (it's allowance of the implementation of efficient generators and engines).

When it comes to disadvantages of the three-phase system, they are many and varied. Among these disadvantages are:

- The necessity of using many expensive wires (three or four)
- The associated large and expensive support systems for the wires
- The need for intermediate stations, sometimes every thirty kilometres
- The very expensive underground and underwater three-phase systems
- The system's strong negative environmental impact
- The frequency of wire breaks
- Large energy losses

The remainder of this book will show that it is possible today to make electrical systems without these disadvantages that still allow for the implementation of three-phase generators and motors.

The three-phase system's deficiencies have contributed in recent years to the development and implementation of high-voltage DC (HVDC) systems.

For very long-distance transmission, HVDC systems may be less expensive and suffer lower electrical losses. For underwater power cables, HVDC avoids the large current requirement needed to charge and discharge the cable capacitance (its ability to hold electricity) during each cycle. For shorter distances, the higher cost of DC conversion equipment compared to an AC system may still be justified, due to other benefits of direct current lines.[2]

The main disadvantages of HVDC are in conversion, switching, control, availability, and maintenance. The required converter stations are expensive and have limited overload capacity.

Operating an HVDC circuit requires the operator to keep on hand a significant number of spare parts, often exclusively for one system. In contrast to AC systems, the realization of multiterminal systems is complex (especially with line-commutated converters); it necessitates the expansion of existing circuits to multiterminal systems. The grounding systems in DC circuits are more complicated and have higher resistance.

Is it time to implement the method first proposed by the great Tesla in 1890?[3]

Nikola Tesla

2   https://en.wikipedia.org/wiki/High-voltage_direct_current

3   See US Patent, publication number 1.119.732, 1890, single-wire electrical energy transmission by Nikola Tesla.

"Transmission of electricity requires at least two wires." This statement has been ingrained in the consciousness of engineers for more than 150 years. How else, after all, could any battery and any coil of a generator have a minimum of two terminals? In addition, any loads for electrical energy have a minimum of two terminals. Therefore, the electrical circuit must have a minimum of two wires. Usually in books, articles, or lectures, authors explain the work of an electrical circuit as the process of current flowing from the generator to the load and then back to the generator. So we need a minimum of two wires.

Nevertheless, the necessity of always having two wires (two channels) is not so obvious (Weber and Nebeker 1994). Actually, during more than a hundred years, humankind has transmitted information and energy from transmitter to receiver by means of the electromagnetic field. Here, we deal with one channel.

Fig. 1.1. Radio communication

If it is necessary to reply over ether, the other channel is used, but once again, it is a single channel **also**. These channels are separated

4

in time or in frequency, or they are distinguished by means of a special code. Somebody might say that, with this kind of radio communication, there are two channels, as the electromagnetic field has two components—electrical and magnetic. Indeed, at the point of reception, there are magnetic and electric fields. Relation between the fields is 120p. Knowing the level of one of these fields and radiation resistance (or current value or effective isotropic aperture) of the receiving antenna, we can compute the active power reaching the receiver. Therefore, we are dealing with a one-way system.

Fig. 1.2. Optical line

The other example of a one-way system is communication by means of fibre-optic lines.

From one end of the globe to another, one optical cable is laid, and it transfers a huge volume of information. No return cable is required. The other cable is used only when it's necessary to transmit additional information or as a backup. It is worth mentioning that the electrical energy for feeding of the optical signal amplifiers, which are being installed over certain distances, is usually transferred over the single wire as well.

Fig. 1.3 Waveguide

One more example of a single-channel system of energy and information transmission is a waveguide, a system that is in wide use today in communication technology.

A waveguide is a channel of either natural or artificial origin that propagates certain waves, such as electromagnetic or sound waves, along an axial line or axial surface with relatively small attenuation. The waveguide limits the existence of the wave by way of the space that is located near this axis or the axial surface (much like a canal limits the waves of the water contained within it).

Let's revert, however, to the wire electrical system – a system that the majority of specialists and amateurs firmly believe should be multiwired. Here, it is necessary to give notice that I am not a physicist and have no intention of giving scientific proof for some new theory of electricity. Here, I only intend to pay attention to some widespread absence of logic in explanations of electrical processes.

Today, many people are not satisfied with widely accepted descriptions of the processes occurring in an electric chain. This description is based on a model in which electrons (or other charges) move inside the conductor. Sometimes it is even presented as the electrons not moving but pushing one another, as in the known "domino effect."

But such an explanation is not plausible. Electrons are mechanical particles, which have some mass. They cannot be moved or push each other with the speed of light. However, the electrical signal is being transferred exactly at the speed of light.

Another contradiction occurs when the operation of the most popular monopole antenna is being described. The current at the end of the monopole should be equal to zero, as it has no way to go any farther. But then this question arises: To where do the electrons that arrive at the end of the antenna disappear? The transmission should be stopped, but the antenna nevertheless operates.

And there is one more contradiction. This one relates to the explanation of processes in grounding systems. In grounding systems, the current enters the earth. At the depth of a few meters, it is impossible to find any traces of this current. Where has it gone? Many attempts to explain these processes have been made. And all of the explanations are different. Some people write that the earth is a huge capacitor. This, however, is not a satisfactory explanation. First, the capacitor should have the second plate. Second, the inside of the capacitor should be dielectric. And the earth cannot be dielectric. Others explain the processes in grounding as current absorption. But absorption cannot be infinite. Any sponge, when it is filled with water, will stop absorbing. Other explanations exist, but all of them give rise to new questions.

However, it is habitual and comfortable to use the term *current*, even if there is no flow.

Let's return to the question of the continuity of the current in an electrical circuit. It is difficult to speak of continuity of the current if the circuit uses a transformer.

Still more difficult to address is the situation in which a capacitor exists, between the plates of which there is a perfect insulator. An insulator does not let current through, but the circuit works properly and in accordance with Ohm's law.

While creating the electromagnetic theory, James Clerk Maxwell introduced the concept of displacement current. In Maxwell's equations, the displacement current, expressed as a quantity, appears in addition to the conductivity current. It is defined in terms of the rate of change of the electric displacement field. Displacement current has the dimension of electric current density, as well as an associated magnetic field, just as the actual currents do. However, it is not an electric current of moving charges but a time-varying electric field. In materials, there is also a contribution from the slight motion of charges bound in atoms—dielectric polarization.

Maxwell conceived the idea and presented it in his 1861 paper "On Physical Lines of Force" in connection with the displacement of electric particles in a dielectric medium. Maxwell added displacement current to the electric current term in Ampère's circuital law. In his 1865 paper, "A Dynamical Theory of the Electromagnetic Field," Maxwell used this amended version of Ampère's circuital law to derive the electromagnetic wave equation. The term *displacement current* is now seen as a crucial addition that completed Maxwell's equations and is necessary to explain many phenomena, particularly the existence of electromagnetic waves.

Thus, we can assume that the current is a mathematical variable equal to V divided by Z. It is convenient for the analysis of electronic circuits.

Thus, one cannot truly conceive of the electric wire as a pipe along which something flows. Such an understanding has likely taken root because of the meaning of the word *current*. Perhaps the concept "current strength" is, in principle, not necessary, even though it is convenient. Still, everyone knows that current strength is the potential difference divided by the resistance, and it may, indeed, be quite sufficient to use these two concepts. Certainly, the concept of current strength is convenient for calculations, simulations, and substantiations of various laws.

This is certainly not the only example of the application of terms or values that do not actually exist in nature but are convenient. For example, in the theory of transformation of signals, the concept of negative frequency is widely applied, even though frequency cannot be negative.

Taking in account these and other contradictions, I have come to the following assumption: Evidently, no "flowing" of the electrical current takes place. The source creates a potential difference. These potentials, owing to the magnetic field existing around the wire, are prorogated along its outer surface with the speed of light. (Hereupon, the well-known skin effect arises). If the potential difference meets a resistance, it produces a work.

It is well known that active energy does not return from load to source. This means that the second channel is not necessary.

True, the second wire may nevertheless be necessary, as it is necessary to transfer not a potential but a potential difference. We'll return to this issue in following section, where it will be shown that energy and information can be transferred on any frequency, including a direct current, over one wire.

There were earlier attempts to perform electrical energy transmission by means of one wire. The Goubau line, or G-line for short, is a type of single-wire transmission line intended for use at UHF and microwave frequencies (Goubau 1950). An AFEP experiment based on a Russian patent application filed on 10 May 1993 by Stanislav and Konstantin Avramenko (PCT/GB93/00960) also attempted a single-wire transmission (Avramenko and Avramenko 1994).

Another straightforward application of the single-wire electrical energy transmission is based on the principle of longitudinal electrostatic waves as described by Nikola Tesla in 1890 (Bettine 2002).[4] These and other known resonance methods use increase in

---

4   Ibid.

frequency. In cases of a large amount of power, these systems must have noticeable losses due to radiation.

So-called single-wire earth return (SWER) systems, which supply low-cost single-phase electrical power from a grid, are now in use and have been for a long time.[5] In such systems, one port of the source and one port of the load are connected to the ground. It will be shown later that these systems have a large amount of reactive power, due to inherent unbalance. This circumstance will be proved in detail.

Assuming that the active energy does not return from the load to the source, we can attempt to construct a single-wire electrical line. The line should not use the ground for the energy transfer from the source to the load. The current, as a matter of fact, will not be entered into the earth. It should not change frequency of source and should not have additional losses as compared with conventional two-wire or three-phase lines.

I am certain that future electrical systems will consist of single-wire lines laid under the earth.

Chapter 2 will present the single-wire balanced electrical system B-Line or single line electricity (SLE) or one-wire system (all of these terms are synonymous) (Bank 2012 and Whitlock 2005).

---

5   See patents WO 2013/018084 A1, US 9246405, US 9419327.

CHAPTER 2

A BASIC UNDERSTANDING OF THE BALANCED SINGLE-WIRE ELECTRIC SYSTEM

A conventional electric A-line (*see* **2.1**, left side) is a combination of the generator and the load connected by two wires, in which phases of currents are opposite (differ by p).

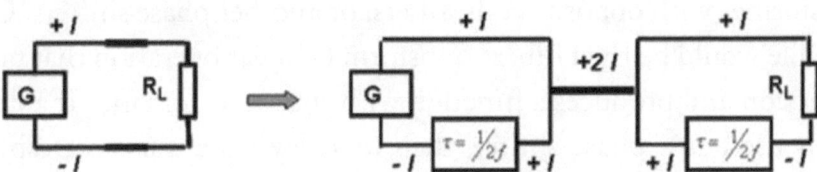

Fig. 2.1. Conventional two-wire circuit
(A-Line) and proposed B-Line circuit.

It is obvious that, if we unite these two wires, we'll get a short circuit. If we nevertheless want them to unite, it is necessary to change the phase in one of these wires to the opposite one. However, the amplitude in both wires should be identical. This can be achieved by inserting a phase shifter in one line.

It is known that the voltage or current phase in an electrical line changes on the opposite one through a time interval equal to half of the period.

For instance, 10 ms (half period) delay line can be used for a signal with the frequency of 50 Hz. Below, it will be shown that on low frequencies, the phase shifter in the form of a transformer with opposing coils is more convenient. After the phase shifter, phases and amplitudes of the currents in both lines are identical, and both lines can be combined into one. However, the load requires a potential difference and not a single potential. Then, in order to provide the normal load functioning, a phase shifter can be inserted before the load in one of the wires. As a result, the two-wire system turns into a one-way B-Line system (*see* **2.1**, right side). That is, the generator and load will "see" the previous conditions (those that existed before the insertion of the phase shifter). In other words, from the point of view of the generator and load, nothing had changed.

The same results can be obtained by inserting phase shifters in both lines and shifting the phase in one line by j and in another line by (p - j). This can be achieved by means of a delay line, a transformer with opposite coils, filters, or another phase shifters. One example would be the Hilbert transform (a linear operation that takes a function and produces a function with the same domain). If a delay line is used as a phase shifter, then its delay time must correspond to the half period. In the case of low frequencies, use of a delay line is practically impossible, since the wire, which corresponds to half a wavelength at 50 Hz, should be 3.000 kilometres long. At low frequencies, it is convenient to use a transformer with reversed coils as a phase shifter. For high frequencies, though, a delay line is a quite suitable solution.

All of the above can be summarized as follows:

1) In a two-wire line, the currents in the wires have opposite phases.

2) The phase of the current in one wire is changed by 180 degrees (π), or the phases of the currents in both wires are changed – in one by plus 90 degrees and in the other by minus 90 degrees.

3) Since amplitude and phase are the same in both wires now, these wires can be combined.

4) The current is split into two currents before the load. The phase in one of the currents is inverted or the phases of both currents are shifted as in item described in item two of this list.

Let's consider the results of simulations at a frequency of 50 Hz.

The proposed idea was checked many times using the ADS-AC program (an experimental open-source program that implements a proposed mechanism for AC called the Absolutely Dynamic System). Series of simulations with different phase shifters and various resistance lines were carried out. Each simulation was carried out for the A-Line and the B-Line. For clarity, the figures below show the conditions and the simulation results, including polarity and magnitude of currents. Figure 2.2 shows one of the simulations for the Ohm's law verification in the proposed circuit. This is the A-Line circuit with the current in the line of 90 mA.

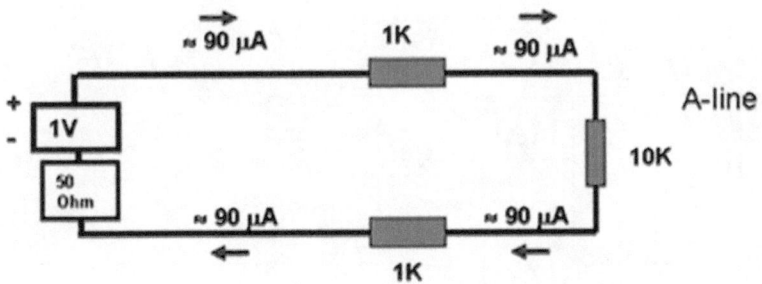

Fig. 2.2. Conventional two-wire circuit
(A-Line) as a prototype for B-Line

In this circuit the resistance of each wire should be equal to 1 kilohm. In the proposed B-Line circuit, we added phase inverters at the input and at the output and combined two lines. As a result, a line resistance became 0.5 kilohm.

The simulation shows that the currents at the input and output have not changed (*see* **2.3**). The polarity of the load current depends on where the inverters are located—at the top or at the bottom.

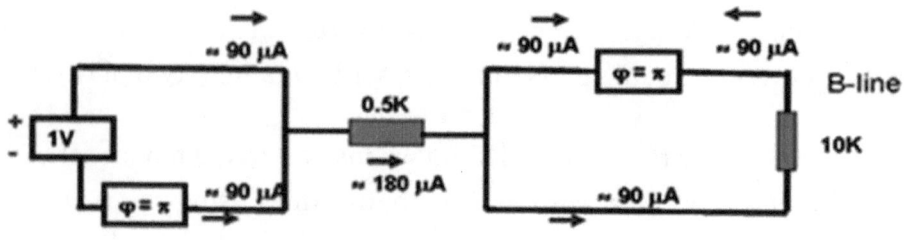

Fig. 2.3. B-Line version of A-Line in Fig. 2.2.

Figure 2.4 shows simulation results of B-Line corresponding to the circuit shown at figure 2.3.

The phase of the current in one wire can be inverted with the help of a transformer with reversed coils The lower ends of the coils should not be connected, otherwise the current will flow from one coil into another.

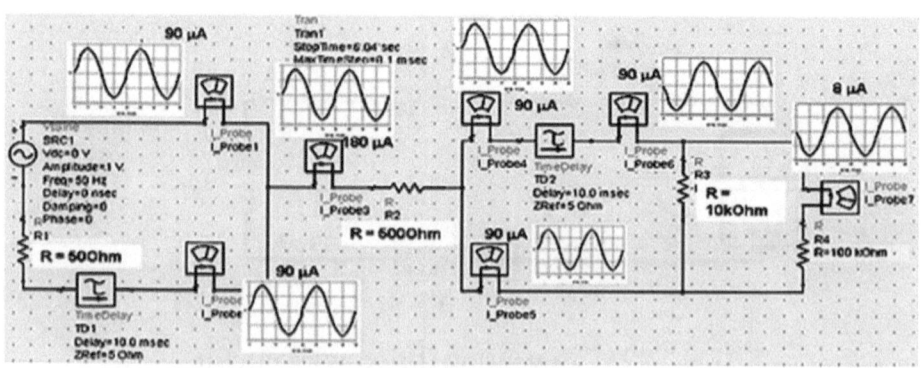

Fig. 2.4. B-Line circuit similar to the circuit
shown at Fig. 2.3 and simulation results.

As it was mentioned above, in practice it is difficult to create a delay line at frequencies of 50 and 60 Hz. But one can apply a transformer with opposite wired coils (*see* **2.5**).

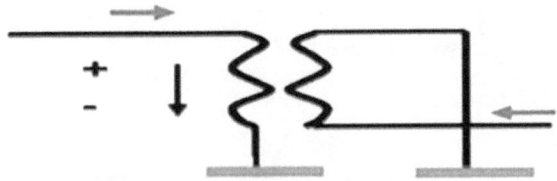

Fig. 2.5. Phase inverting in one line by means of a transformer.

Such a phase inverter requires a zeroing of the middle point potential, around which phase turns takes place. In other similar cases, zeroing is usually performed with the help of the grounding.

As it will be shown below, the grounding in this case does not take part in the process of energy transmission from the generator to the load. Moreover, it will be shown that the zeroing can be performed without the current entering into the earth.

The simulation results in a circuit with ideal 1:1 transformers, as shown in figure 2.6.

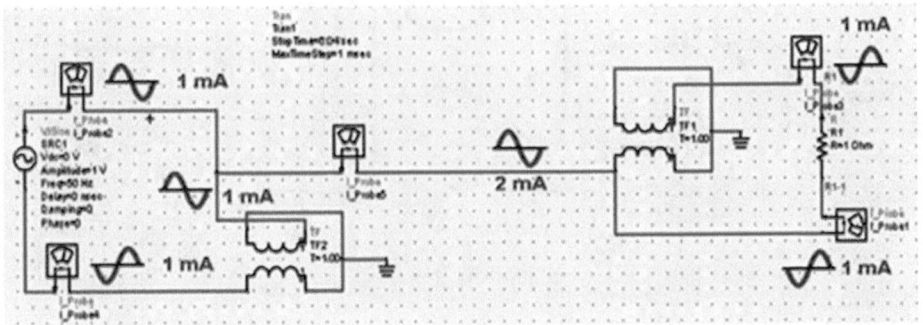

Fig. 2.6. B-Line in which transformers are used as inverters.

In this circuit, the current in the common wire corresponds to Ohm's law. Therefore, no other current (for example, in the ground) can be.

If the B-Line is used in a system with an increasing or decreasing voltage, the transformers must be used at the beginning and at the end of the line. An example of such a system with voltage in the line 6 kV is described in chapter 6, "Experimental Systems" (Bank 2012 and Whitlock 2005).

DC B-Line

Implementing the inverter (phase shifter) in a DC circuitry requires a different solution than the aforementioned transformers. According to the primary idea of the B-Line, it is proposed to use two capacitors and corresponding switches to implement the inverter as shown with respect to figure 2.7 in the source side and, correspondingly, at the load side.

In this circuit, we will designate as *period A* "the state when the keys are closed in point A." *Period B* will be "the state when the keys are closed in point B."

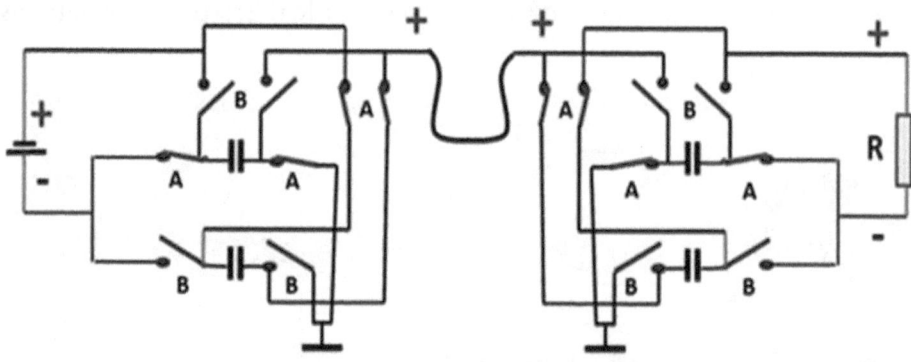

Fig. 2.7. DC B-Line example

Each of the inverters operates as follows: In period A, the first capacitor is charged, and the second is discharged. In period B, the two capacitors reverse functions. The charge current goes in one direction, and the discharge current travels in the reverse direction.

As a result, the sum current in the common wire will have the polarity of the current, which does not take part in the charging of the capacitors.

In this example, the in-line current has one direction, positive or negative. In figure 2.7, the direction is positive.

One can set the charging and discharging times of the capacitors, taking into account the value of load resistance. Such a DC B-Line system can be implemented in an electrical railway system (in other words, a tram). In this case, it is possible to transmit electrical power only in the wire or only in the rails.

In the circuit shown in figure 2.7, a serious problem connected with zeroing of potentials exists. This problem is caused by the greater difficultly posed by the need to ground direct current than to do so for alternating current – an issue that is well known among those who work with systems of electric transportation and single-wire power supply systems of amplifiers for optical lines.

However, even for this case, an acceptable solution has been found. That solution will be described in chapter 5.

Here at the conclusion of this chapter, it is necessary to notice that the current in a single-wire system is twice that in a two-wire system. This means that wire resistance should be two times lower. However, one wire with linear resistance R^2 is cheaper than two wires with linear resistance R. In addition, the transmission towers will cost much less. Possibly the biggest advantage of a single-wire system is its capacity for cost-effective energy transfer underground or under water (see more on this in chapter 9).

CHAPTER 3

THE NULLIFER: ZEROING WITHOUT INSERTING CURRENT INTO THE GROUND

An electrical ground system should have an appropriate current-carrying capability in order to serve as an adequate zero-voltage reference level. In electric circuit theory, a "ground" is usually idealized as an unlimited current source or the charge absorber, which can absorb an unlimited amount of current without changing its potential (Bank 1982).

Fig. 3.1

The linear resistance of ground (between two points on the surface) is great (50 to 1.000 **100** ohms per metre). Therefore, energy cannot be transmitted between two points through the ground.

Potential zeroing is the main objective of the grounding in the systems for electric power transmission over considerable distances. As is shown in various sources, if grounding properly executed, it is impossible to discover any current traces at depths of greater than ten metres (Gerke and Kimmel 1998, Morrison and Lewis 1990, and Bank 1982). (see reference)

However, in many simulation programs, such as the ADS program, there is only one ground bus. Therefore, it is impossible to separate the transmitting and receiving parts of the system, since they have joint ground (common grounding).

Grounding in B-Line is intended for potential zeroing only. However, zeroing can be accomplished by other methods as well, without using a ground. For example, if two B-lines with opposite phases are remote from a source, you can combine both points designed for zero potential. In this case, you don't need grounding as shown in chapter 6. It shall be noted that this method is quite acceptable in practice, as high power stations often have several outputs.

However, in almost all electric systems, there exists the necessity of potentials zeroing. This can take the form of grounding for protection systems or zeroing of the neutral line in a three-phase system. In addition, potentials can be zeroed in an asymmetrical antenna, for example, a monopole, or in a single-wire system, for example, SWER or SLE.

The zeroing is also necessary in some systems that work with direct current. Examples include electrical transport facilities or power supply systems for optical cable amplifiers.

Today, zeroing is made by means of grounding. Usually, grounding is achieved via a metal pin, about 1.5 or more metres in length, inside the ground.

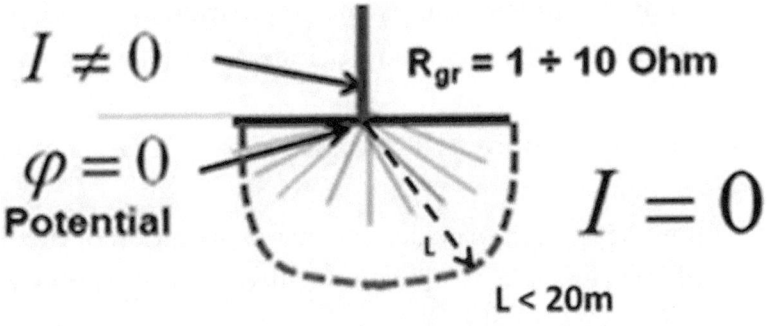

Fig 3.2

If resistance of grounding results in unacceptably high energy losses, one can use several groundings, which are connected in parallel and located at distances not less than ten metres between them (*see* **3.2**).

One can measure grounding resistance by giving currents to two groundings. The resistance of the two grounding does not depend on the distance between them, if it is greater than about ten metres (*see* **3.3**).

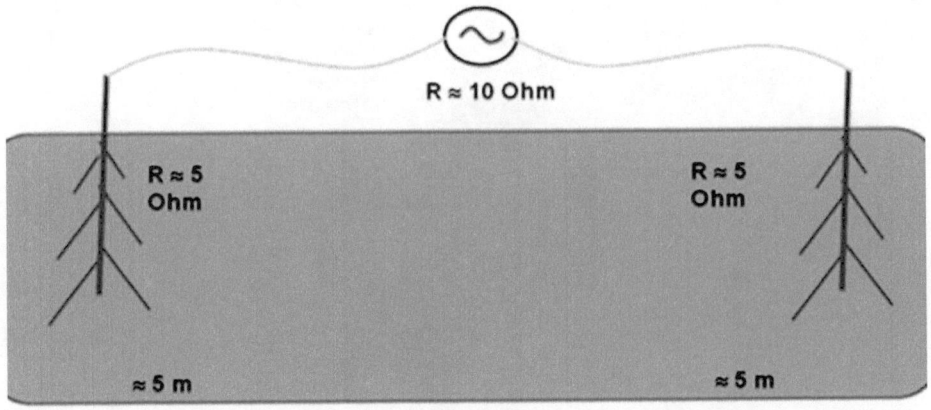

Fig. 3.3

Typically, grounding creates an array of problems. Many of them are connected with the instability of ground parameters in relation to changing weather conditions. In addition, grounding often requires a large area. On top of that, in some countries, it is possible to use grounding only for protection, as the population is afraid that the electric current can harm animate beings as well as plants in the ground.

Today, there are no universally recognized paradigms for grounding operations.

The following section provides another explanation of the zeroing process and officers a new device—called a "nullifier"—for the realization of zeroing.

AC zeroing system: An antenna

The current that is being injected into the ground is divided into a great number of weaker currents. When the ground depth increases, the number of currents grows; hence, the amplitude of each current decreases to zero.

Let's consider the protection grounding of an electrical cabinet.

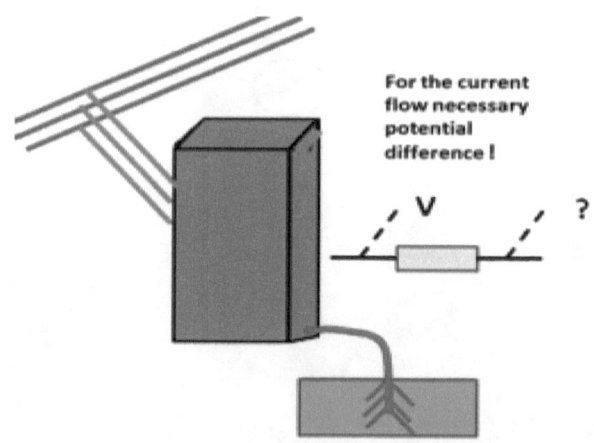

Fig. 3.4

Potential difference is necessary for the current to flow. But we have here only one potential (V), not a difference of potentials like that found in a broken wire. Can a current flow through broken wires? Yes, it can; for example, it does so in a linear antenna.

In the case of an electrical antenna, like a dipole or monopole, the current stops at the ends of radiators, but its energy is converted into the energy of the electromagnetic field.

This means the energy path is not interrupted

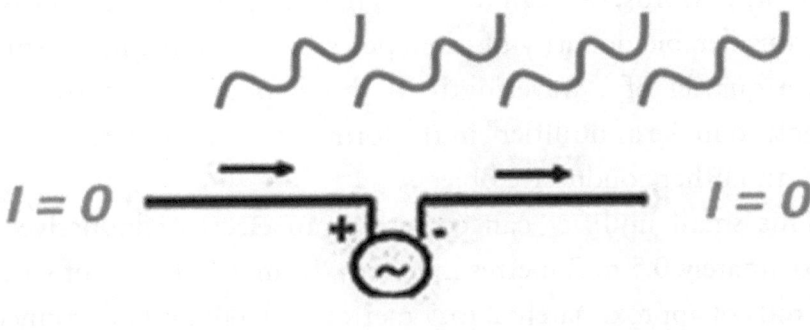

$$I = 0 \qquad\qquad I = 0$$

Fig 3.5

Now we can imagine a lot of ground connected at the input of very short monopoles.

It is known that a monopole with height (h), where h << l/4 has radiation resistance, equals (Bank 2014):

$$R_{rad} \approx 14(mh)^2 \Omega, \qquad m = \frac{2\pi}{\lambda}$$

This resistance tends to become zero with a decrease of h compared to the l/4/. Decreasing radiation resistance leads to decreasing of radiating power because

$$P_{rad} = R_{rad} \cdot I^2_{rms}$$

So we can say that a monopole with h 5 – 10 m at frequencies 50 or 60 Hz has zero resistance and zero radiation field density.

A monopole with a height of much less than a quarter of the wavelength has a capacitive component (Xue et al. 2012). However, parallel connection of monopoles also results in the decrease of capacitive resistance.

In other words, one can tell that grounding is an aerial consisting of a considerable quantity of monopoles, with lengths much smaller than a quarter of a wavelength. If this hypothesis is correct, it is possible to make a "nullifier" in the form of a device isolated from the earth and other conductive objects.

This small nullifier can consist of an electro conductive rod, approximately 0.5 to 2 metres in length, from which a set of wires or thin rods of approximately 2 to 5 metres stick out and are connected to the central pin. The supports should be made from non-conductive material. An example of the nullifier, where electrical wires are used as a line, is shown in figure 3.6.

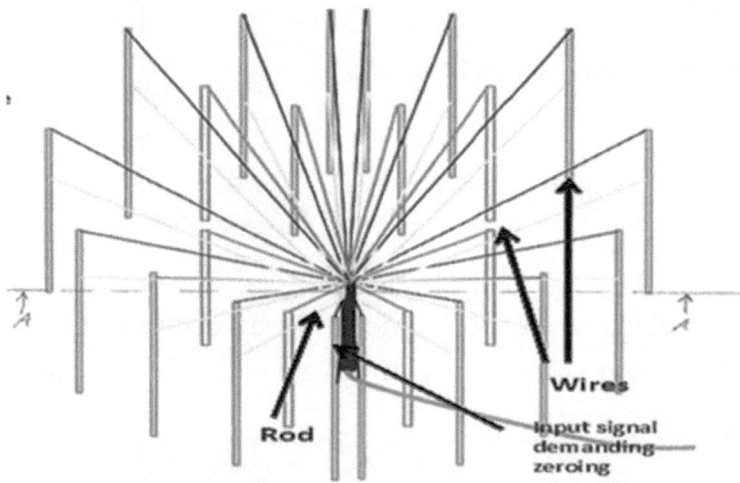

Fig. 3.6

One can make a nullifier by taking the isolated volume (*see* **3.7**) filled with a semiconducting material similar to soil or soil itself. It can be a concrete pipe of round or rectangular cross sections from 3 to

10 metres long and cross sections of 10 to 15 square metres. As in the case of usual grounding, the conductive rod (about 1.5 metres long) is inserted into this volume.

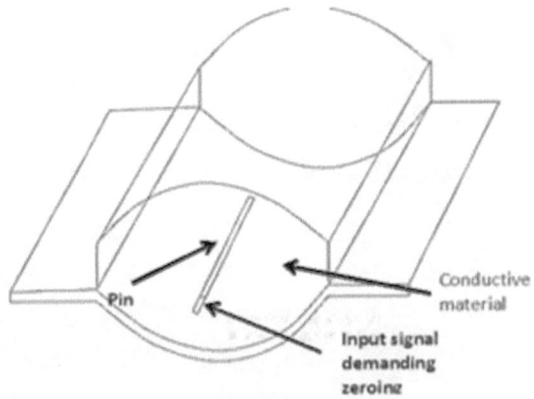

Fig. 3.7. Nullifier structure

The real nullifier of such type is shown in figure 3.8.

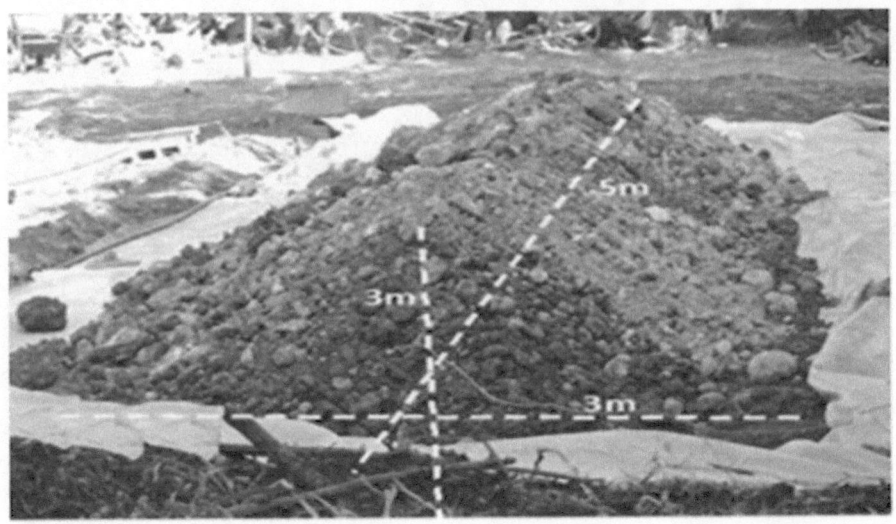

Fig. 3.8 Real small nullifier.

For implementation of such a nullifier in an experimental, single-wire 6 kV system, a ditch of 3 metres in depth and 5 metres in length has been dug. The walls and bottom of the ditch have been covered by a non-conductive nylon type fabric. The results of this nullifier implementation are given in chapter 6.

The needed resistance can be provided by a large nullifier. This experimental nullifier was built in Tal Shahar, Israel, as a concrete structure in the form of a pentagon. The diameter of the inscribed circle is 10 metres, its depth is 3 metres, and its concrete walls and floors are 12 centimetres thick (*see* **3.9**).

Fig. 3.9.

The container was filled with soil. The top point of the ground was a metre above the concrete wall of the nullifier (*see* **3.10**).

Fig. 3.10

Four rods were then placed between the edge and the centre of the nullifier. Thus, with the use of five parallel rods, grounding resistance equal to 4 ohms was obtained.

In addition, the capacitance between the nullifier's electrode and the external grounded electrode was measured. The capacitance was several nanofarads. Therefore, at 50 Hz, the resistance of this capacitance equals about 3 ohms

The same idea of the nullifier resistance decrease can be applied in case of "air" nullifier usage shown in figure 3.5. Then we get the design shown in figure 3.6. If the number of wires used to make the nullifier (*see* **3.11**) wasn't sufficient or the volume of the nullifier (*see* **3.6**) wasn't large enough, the nullifier's capacitance resistance will be too large. In this case, the capacitance resistance can be compensated by way of the incorporation of an inductance L between the point, which requires the zeroing and the nullifier as shown in figure 3.11.

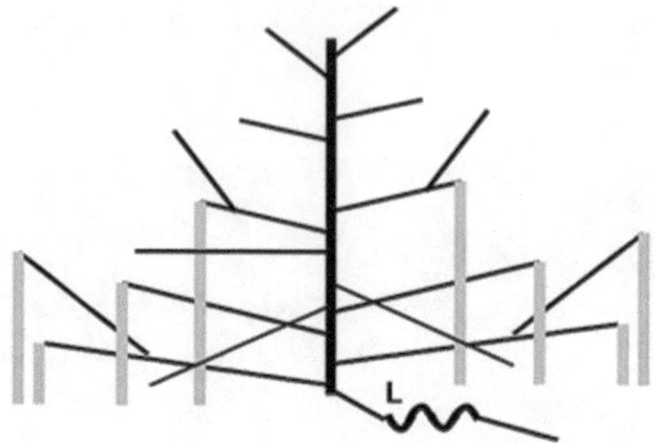

Fig. 3.11. The idea of an air nullifier

This method has been checked at the frequency of 1 MHz. (The programs that we know don't allow for antenna simulation on a frequency of 50 Hz.)

At first, the monopole (with a height of 12 millimetres, much less than a quarter of the wavelength) had been checked (*see* **3.12**).

Fig. 3.12. Short monopole

In this case, active resistance will be 0.65 ohm and reactance will be 453 ohms. Then the additional "rays" were added *(see* **3.13**).

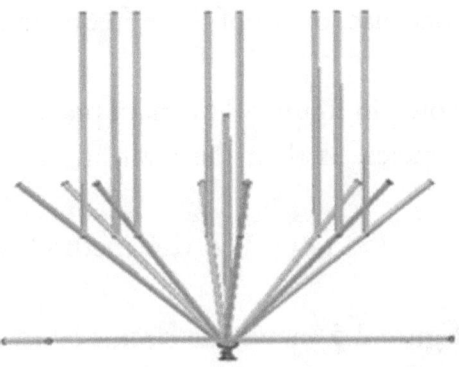

Fig. 3.13. Monopole with additional rays

In this case, the active resistance was equal to 0.8 ohm, and reactance was 60 ohms. Then between the source and the input of the "antenna," an inductance was incorporated. In this case the active resistance has proved to be 1.0 ohm, and the reactance, 1.98 ohms. Thus, such a circuit is completely suitable for the zeroing of potential.

Single frequency nullifier

This section proposes a zeroing using a device that does not also contain a ground, in order to resolve this problem in simulations. Its functioning is clearly shown in figure. 3.14.

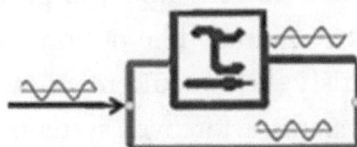

Fig. 3.14. Signal zeroing at single frequency

Here zeroing takes place by the addition of two currents of the same amplitude but with the opposite phases.

This method of zeroing (or this nullifier) works only at frequency $F_o = 1 / 2t$ or at all other frequencies satisfying the condition $F_n = 1$

/ (2n + 1). This is because, on all of these frequencies, the delay gives phase shift 180°.

The method shown in figure 3.14 can, in practice, be implemented only at high frequencies, as then the realization of the delay line is not a problem.

At low frequencies (50 to 60 Hz) this method is suitable only for simulations.

In chapter 2, figure 2.5 shows the circuit of an inverter intended for transformation of a two-wire signal into a single-wire signal, and the requirements for zeroing are given in the middle points. The application of a nullifier, which is described here, allows for zeroing without current injection into the ground (*see* **3.15**).

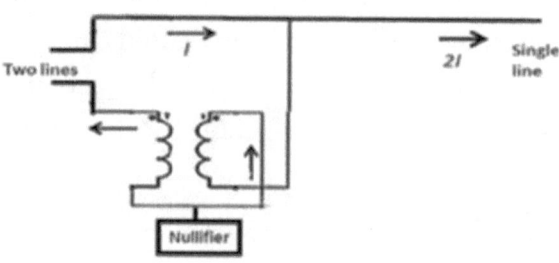

Fig. 3.15. Inverter with single-frequency nullifier

As noted above, many countries (at the request of the "greens") only allow injection of a current into the ground (at systems) for protection against short circuits. The options presented here in the form of the nullifier satisfy these requirements without decreasing the efficiency of electric systems. Moreover, systems of zeroing that don't require grounding into the earth have the important advantage of not depending on weather conditions.

Let's consider the effect of the nullifier's grounding resistance by exploring an example. We'll assume that it is necessary to supply 2.2 megawatts (MW) of electrical power in a city. The main voltage in the city is 220 volts (V). Thus, the consumption current is 10.000 ampere (A), and the load resistance is 0.02 ohm.

We'll supply electrical energy with voltage of 220 kilovolts (kV). That means a step-up transformer will "see" $0.02 \cdot 10^6 = 20$ kilohm. (Today kOhm is customary)

In order to get a one-wire line, let's apply an inverter as shown in figure 3.15.

However, this circuit will not operate because both currents go towards each other and create an infinite resistance.

In order to get a single-wire line that actually operates, let's apply an ideal inverter—in other words, the inverter, in which the wire's resistance is equal to zero and the grounding or zeroing is ideal. Now the currents flow, and there are no losses.

Let the resistance of the grounding or nullifier be equal to 10 ohms; this means the useful current will flow not only through the 20 kilohm but also through the resistance 10 ohm. It is obvious that the losses will be negligible.

One can believe that the zero point of grounding is a point inside of ground, where the current is equal to zero and the resistance of grounding is negligible.

The proposed nullifier operates quite in the same way. The current in the antenna end is always equal to zero, and the radiation resistance (impedance) of a short (as compared with wavelength) antenna is very close to zero. The "air" nullifier can have a capacitive resistance. But first, it can be compensated, like in figure 3.11.

The aforementioned arrangement allows for the suggestion that the air nullifier can be grounded by an antenna consisting of a great number of shorter monopoles. The ground is a conducting granular substance. Currents flow between granules going over the distance of several meters. One can envision this as very short monopoles (a quarter of the wavelength at a frequency of 50 Hz is 3.000 km). That means that these monopoles have an active radiation resistance equal to zero and a large capacitive resistance. But because there are many such monopoles, the capacitive resistance is very small too. Such an antenna radiates practically nothing.

One may ask why currents flow in ground not deeper than ten metres. This phenomenon can be explained as follows. Let's assume we applied a potential to a usual ground. At the first moment, both the length and the quantity of the currents will be small. In this, case the capacitive component of the grounding impedance will be large, and there will be no zeroing of potential. Soon comes the moment when the grounding resistance will be close to zero. The zeroing of potential begins at this moment, and the increasing of the length and quantity of currents therefore ceases.

Grounding in DC systems

The problems created by currents flowing directly into ground, include:

- Electrochemical corrosion of the long metal objects inserted into the ground (such as pipelines)
- The production of chlorine or other affect on water as a result of the use of seawater as the second conductor (and, thus, the current flowing into the seawater)
- The appearance of a magnetic field as a result of the current in water, which can affect the performance of the magnetic navigation compasses of ships sailing over the underwater cable

Another serious problem is the wandering currents. The sources of wandering direct currents are typically electric trains, groundings of direct current lines, installations for electric welding, systems of catholic protection, and installations for the deposition of galvanic coverings. An example of an appearance of a wandering direct current is a tramline, where steel rails are used for second current return to a generating station flow.

Owing to bad contact between rail joints and insufficient isolation from the ground, a part of the current enters into the soil and finds ways through objects of low resistance, for example, through underground gas pipes and water pipes. If the pipe is protected by a non-metallic covering, it aggravates the corrosive destruction, because, in this case, all wandering currents leak through defects in the pipe coating. This, in turn, causes the current density to increase in the areas with limited surface and accelerates pipe destruction.

The grounding of a direct current transmission line requires a complicated and labour-intensive installation. The installer must create reliable and constant contact with the ground in order to net the desired result and to eliminate the possibility of dangerous "step voltage" occurring.[6]

Examples of the application of single-wire DC systems include the power supply systems of optical cable amplifiers, including the amplifiers for underwater lines. Zeroing in on coastal zones, installers often drill deep boreholes to reach groundwater. Next, rods with good electric conductance are inserted into the boreholes. More difficult circuits exist as well. The example given in the source cited in note 5 shows that, for a current of 225 A, it is necessary to use a rod of 33.54 metres in length.

Various sources indicate that resistance of grounding for alternating current can be of the order of 10 ohms.

---

6   One of many sources that document that grounding a direct current transmission line is a complicated, labour-intensive process can be found at < http://forca.ru/stati/energetika/osnovnye-nedostatki-setey-vysokogo-napryazheniya-postoyannogo-toka.html> (Russian).

The zeroing system for DC

No antennae works on direct current. But groundings are necessary for the zeroing of potential, as well as for static electricity removal. In the presence of grounding, static electricity charges are led into ground and, thus, don't accumulate to a magnitude at which sparks would be possible. The grounding device consists of steel pipes of between two and three metres in length. These pipes are dug into the earth in such a way that their top ends are half a metre below the earth's surface. The pipes are connected to the surfaces of protected objects using steel strips – devices, cars, and pipelines that require grounding.

The following hypothesis can be made from the aforementioned data and a variety of other sources: For DC potential zeroing, it is possible to get rid of static electricity by connecting the necessary point to the large metal object. A cylinder (pipe) made of conducting material of a certain length and thickness can obviously be such a device.

To estimate the possible sizes of cylinders, the following simulation has been performed: A cylinder, which is located above a conducting surface, as is shown in figure 3.16, was taken.

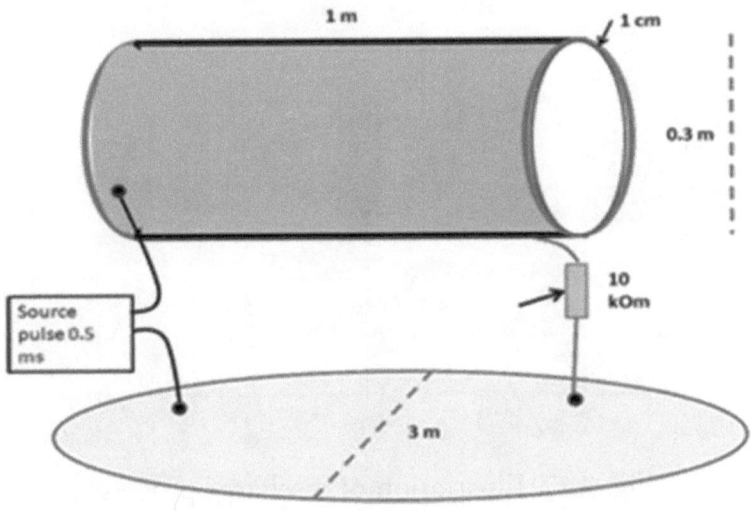

Fig. 3.16

The impulse with the duration of half a millisecond was used as an input signal. At the end of the impulse, the current flowing via resistor 10 kilohm was measured. It can be suggested that if, at the end of the impulse, the current is not equal to zero; the potential (which is an equivalent of static electricity) remains on the cylinder. During the simulation, the different sizes of the cylinder, which made it possible to decrease the magnitude of the current, were being chosen. If the magnitude of the current is less, it means that a part of the potential has discharged. Some results of the simulation are shown in figure 3.17.

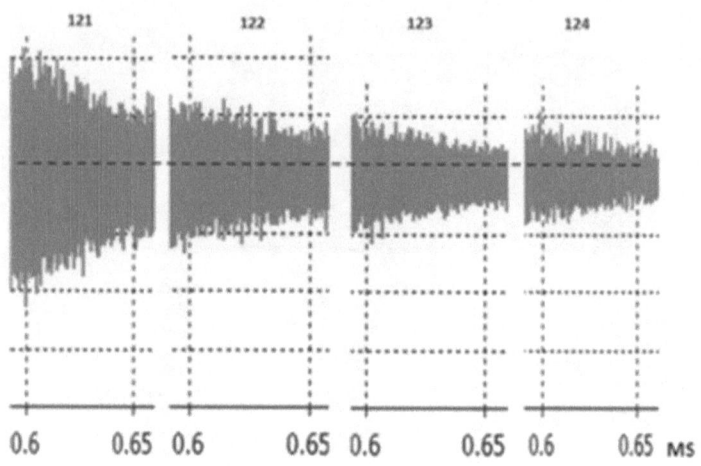

Fig. 3.17. Illustration of discharge process

This simulation includes a repetition of the above process.

Taking into account the results of the simulation, one can suggest that an acceptable zeroing can be achieved by means of one or several cylinders one to three metres in length and wall thickness of ten to thirty centimetres. The material of the walls must not have very high conductivity. It can be aluminium or a cylinder inserted in a box filled with grounding material.

The rod, to which constant voltage is applied, represents the cylinder over which a surface a charge is regularly distributed. This charge possesses potential energy. If all points at the cylinder's surface connect using resistors to points located at a great distance from the cylinder where electric potentials are equal to zero, the current will begin to flow through the resistors. If parallel, the resistance of all the resistors will be close to zero, and the electric potential of the cylinder will be close to zero as well.

In our case, the "ground" in which the zeroing rod is inserted is all these resistors. As resistance of the ground can be very big, the chosen rod should be long enough. This rod is many times longer than the rod being used in AC systems. That is why, when comparing AC and

DC systems, the complexity of manufacturing and operating a system for zeroing is the important factor in favour of AC systems.

Different sources indicate that the resistance of the grounding of direct current may be of the order of 100 ohms.

CHAPTER 4

BALANCED AND UNBALANCED SINGLE-WIRE SYSTEMS: THE SWER SYSTEM

Unfortunately, the words *balanced* and *unbalanced* are used in two different cases. First, the transmission lines can be either balanced or unbalanced with respect to zero potential. Second, a three-phase system is balanced or unbalanced with respect to whether or not equality exists between loads.

Considering this, it's important to note that, hereinafter the words *balanced* and *unbalanced* will be used for first cases. For the second case, we will use the phrases *load balanced* and *load unbalanced*.

Figure 4.1 illustrates the normal construction of balanced and unbalanced lines. Balanced lines are used more for transmitting information signals. This allows decreasing the noise influence and interferences (Gerke and Kimmel 1998).

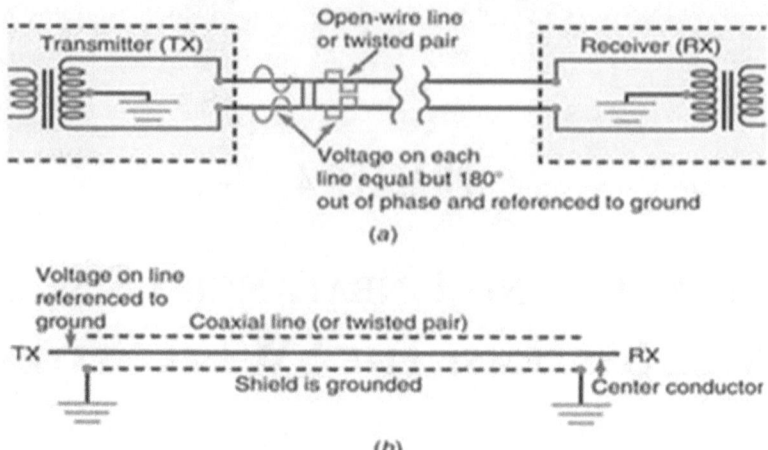

Fig. 4.1. Balanced line (a) and unbalanced line (b)

In an electronic and electrical circuit theory, a "ground" is usually ideal sink or source of an infinite amount of charge, which can absorb an unlimited amount of current with potential of the grounding point equal to zero (Gerke and Kimmel 1998 and Morrison and Lewis 1990). For more details about grounding, refer back to chapter 3, where it is shown that energy cannot be transmitted between two points through the ground. However, grounding can help transmit energy through a single-wire system.

All three circuits shown in figure 4.2 (a, b, and c) are equivalent, if the distance between a source and a load is much less than the wavelength. In practice, if two points of the circuit, for example 3.2 a, have identical (zero) potential, these points can be connected (*see* **4.2b**).

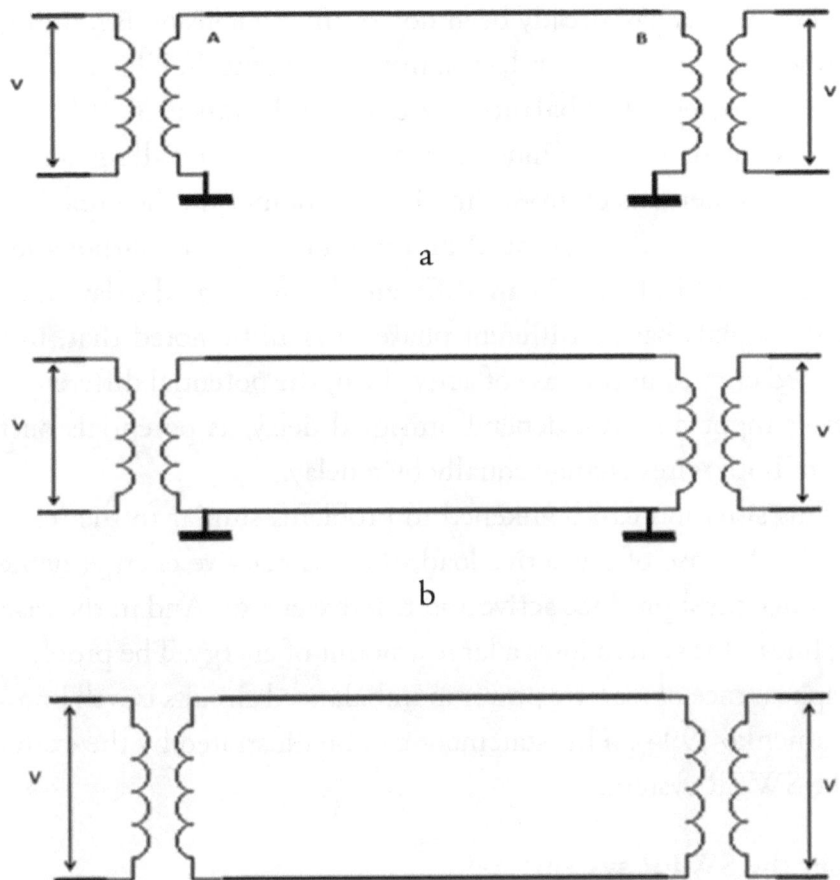

Fig. 4.2. Equivalence of balanced and unbalanced circuits

By not using both zeroings (groundings) in the circuit in figure 4.2b, the magnitudes of the currents will not change. In the circuit shown in figure 4.3d, in both wires, there are currents that correspond to Ohm's law. In the circuit shown in figure 3.2a and 3.2b, the current is only in one wire. But in these cases, potential in the working line is above the zero level. Therefore, the load will receive the same power as in the case of the circuit shown in figure 3.2c.

In other words, from the point of view of energy transfer, the two-wire balanced circuit and the single-wire circuit of a SWEP type are equivalent.

However, as has already been noted, this conclusion is fair only if the lines lengths are much less than the wavelength. This fact is the primary problem of unbalanced systems, for instance, SWER.

This unbalanced line must transmit all energy and all signals from Tx to Rx. Potentials of grounding in both points are the same (zero). However, potentials in points A and B (on transmitter output and on receiver input) in Fig. 3.2a are different due to a signal delay, and Tx and Rx signals have a different phase. It shall be noted that, in the balanced circuit, in the case of active load, the potential difference on receiver input does not depend on signal delay, as potentials on the ends of both wires change equally by a delay.

This situation can be likened to problems similar to the reactive load. In the case of a reactive load, there is reactive energy; namely, the source must produce active and reactive energy. And in the case of long lines, this system loses a large amount of energy. The problem of the appearance of reactive power in unbalanced circuits is well known. (Avramenko 1994). This statement can be illustrated by the example of the SWER system.

About the SWER system

The authors of the single-wire earth return (SWER) system claim that SWER employs only one conductor and that the current return path through earth (*see* **4.3**) is employed as an electricity distribution method.[7]

---

7   See patents WO 2013/018084 A1, US 9246405, US 9419327.

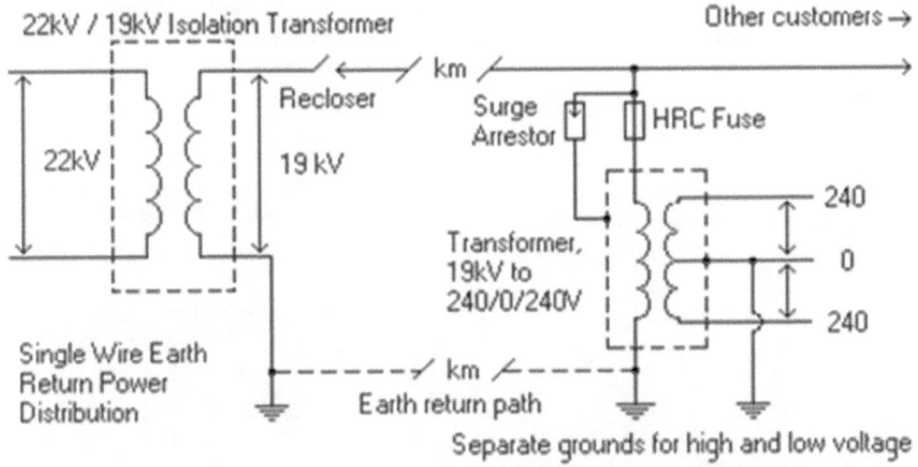

Fig. 4.3. The circuit of the SWER system

Emergence of reactive power in a real SWER system can be shown with the help of a simulation (*see* **4.4a**). In this case, the phase shift in a line is cos j = 0.65, which corresponds to a line length of 450 kilometres. It is equivalent to the load shunting by a reactance, which must result in an increase of the current of generator. In this case, the generator current has appeared more than twice higher than that in the load (*see* **3.4b**).

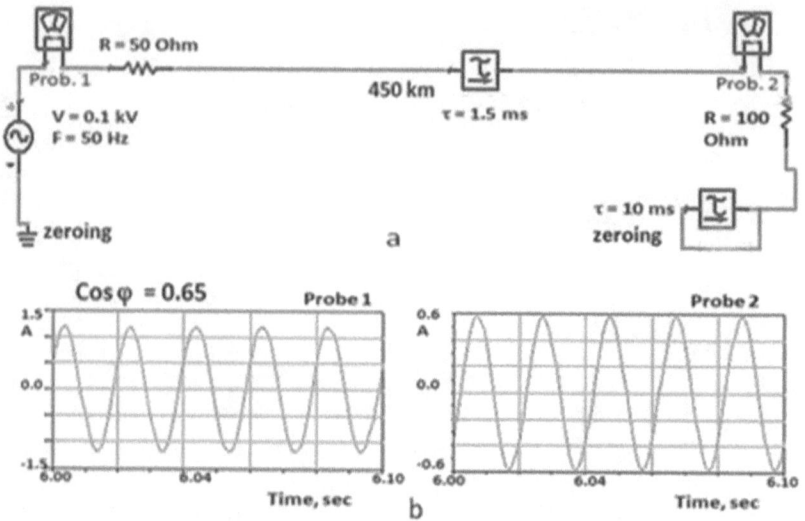

Fig. 4.4. SWER simulation circuit (a) and simulation results (b)

Compensation of the reactance, which emerges owing to misbalance, creates problems due to both instability of the load and possible losses in compensating filters.

Figure 4.3 illustrates that the SWER system is usually an unbalanced circuit. Therefore, some losses of energy can be expected. However, these losses are not caused directly by the fact, that the low points of transformers are connected to the earth. On the contrary, as shown above, the zeroing leads to losses.

Here, it is clear that the generator must produce a current that is greater than the current in the load. This problem cannot be solved by means of compensation using series or parallel inductivities or capacitors. That is because the phase shift of these filters depends on the value of the resistive load, which, typically, is not constant. These filters have significant losses, which considerably complicates the implementation of the phase compensating circuit.

Balanced one wire (B-Line) system

The author proposes a new single-wire system (B-Line)[8]. The single-line method operates as follows: Phase-shifting devices in both wires of a two-wire line or in three wires of a three-phase line set equal magnitude of the phase in all wires. Therefore, the wires can be combined. Prior to loading a single-wire split into two or three wires, they restore the required phases. The generator in the B-Line circuit produces exactly the same two currents as in the usual balanced two-wire circuit. For this reason, the B-Line circuit can be considered balanced. Figure 4.5 illustrates the circuit and results of the simulation of a B-Line system for comparison with the results of the SWER system (*see* **4.4**).

In the case of B-Line, the current of the generator is almost equal to the current in the load.

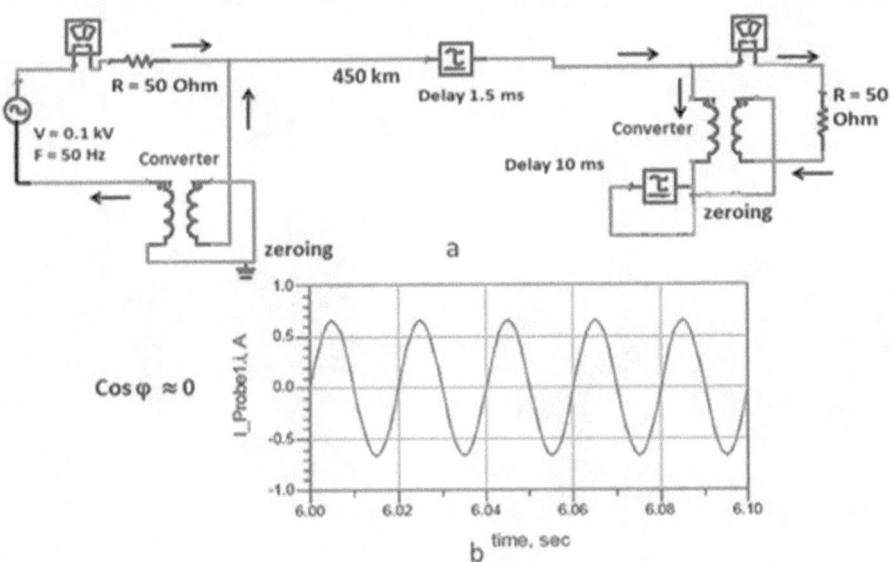

Fig. 4.5. (a) Line simulation circuit and (b) simulation results

---

8   US Patent 1,119,736, 1890, Tesla.

The selected parameters of the delay line correspond to the required phase shift of the load current. The main advantage of a B-Line circuit is its ability to transfer a differential (balanced) signal by a single wire.

CHAPTER 5

THREE-PHASE SYSTEM BY A SINGLE WIRE

The well-known three-phase system is a combination of three single-wire systems. If all three phases have the same load (a balanced circuit), the current in the common wire will be zero. In this case, the fourth wire is not necessary. A part of problems arise if phase loads change unequally. The main advantage of three-phase systems is their use of three or four wires instead of six lines for transmitting three signals, since a three-phase signal is better for some electrical motors and generators.

The disadvantages of the three-phase system include the necessity to use three or four wires to transmitting one signal for a three-phase load, the large distance between wires in many cases, the necessity to use intermediate stations, the more expensive transmission line towers, and a higher line voltage, which is 1.73 times greater than the phase voltage.

We can demonstrate that the signal being transferred by a three-phase system can be transmitted over one wire. Unfortunately, changing a phase by 120 degrees is not a simple task. A simple one-level filter can change phase by less than 90 degrees. The usage of more complex filters in power systems is complicated and can cause significant losses.

However, the task of transforming a three-phase signal into a single-wire signal can be achieved by means of a simpler method, which I will offer here. The solution is a converter application. How a convert application can be employed to accomplish this task is shown in figure 5.1a.

In this case, the phases of two signals are shifted by 60 degrees by means of simple filters. The phase of the third signal is changed by 180 degrees with the help of an inverter.

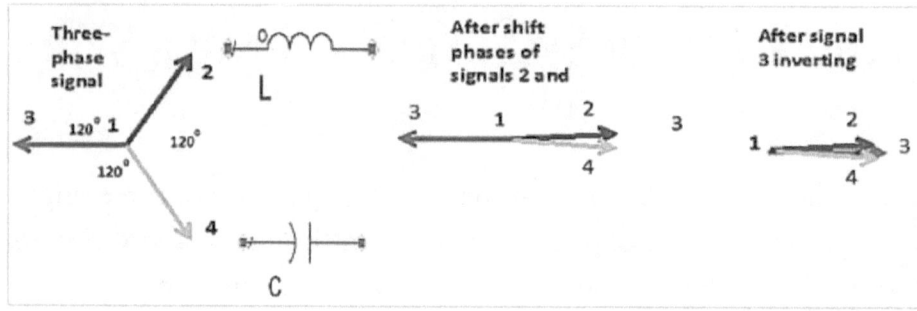

a

The same method in opposite order allows for the transformation of a single-wire signal to a three-phase signal (*see* **5.1b**).

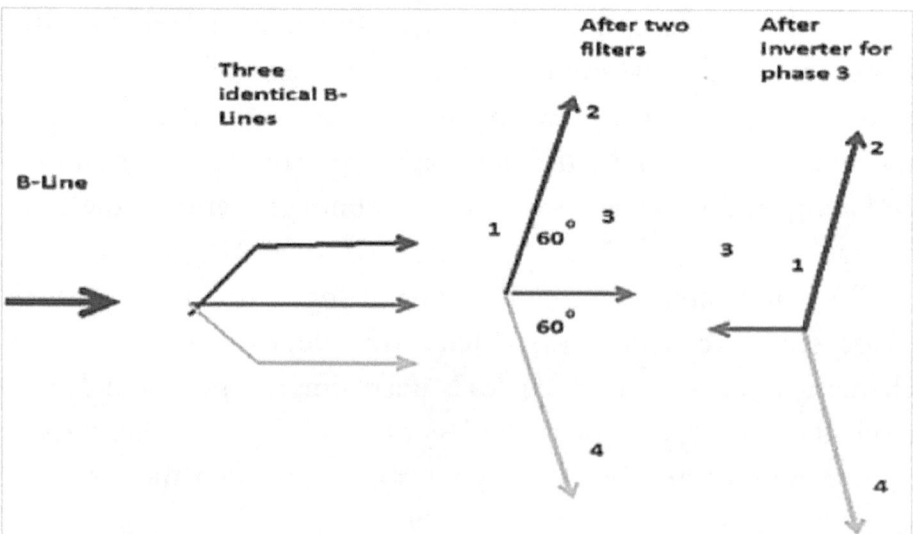

b

Fig. 5.1. The vector diagrams: three to one (a) and one to three (b)

The inverter here is a 1:1 or 1:2 ratio transformer, with the opposite connected windings. All three elements are determined by the given voltage and current.

The values of C and L, together with load resistance R, must provide the phase shift equal to 60 degrees. Therefore, the ratio of Xc to R must be 1.73. The resistance R can be found from the known full load power.

As a result, the converter circuit looks as is shown in figure 5.2.

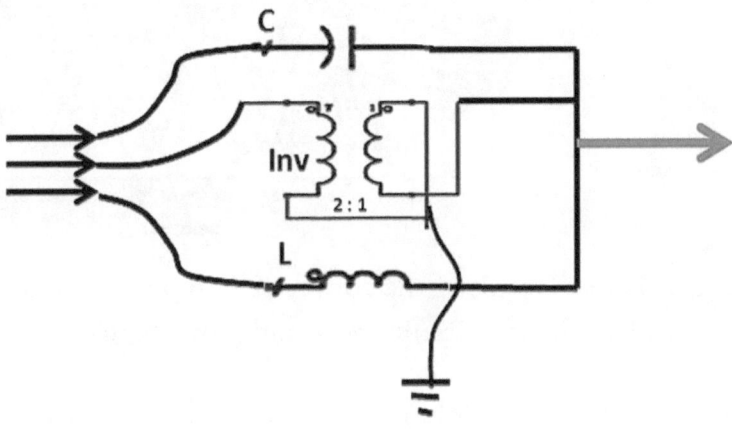

Fig. 5.2

The circuit and the simulation results of the proposed converter are shown in figure 5.3.

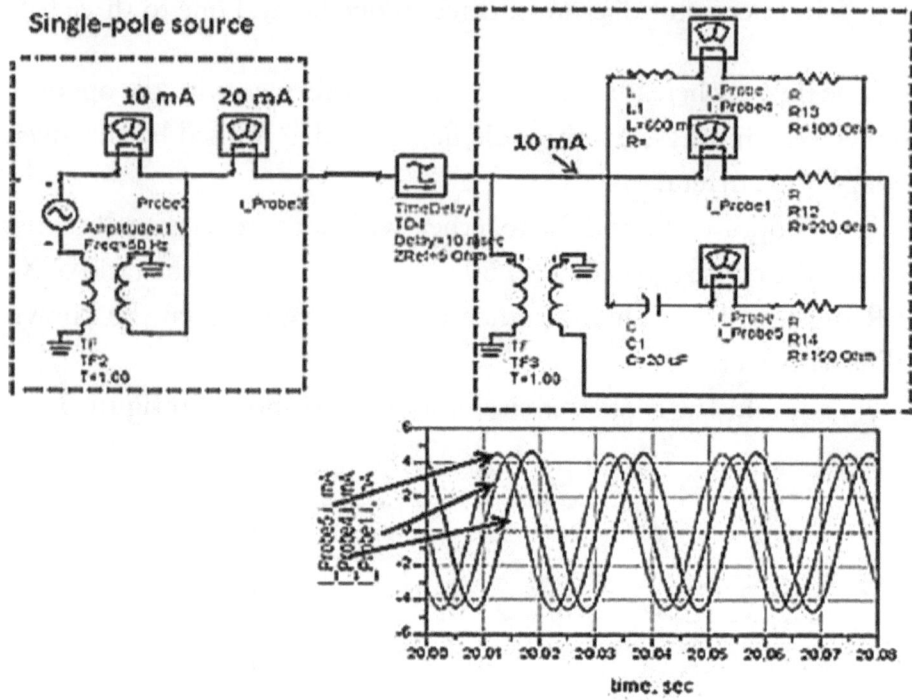

Fig. 5.3. The circuit of the converter 1 to 3 for a three-phase B-line circuit and simulation results

Note that this circuit does not need an additional wire, even in the case of three different load resistances (phase unbalanced circuit).

In practice, the load in the system can change. Therefore, it is necessary to change the values of capacity C and inductance L accordingly.

These changes can be made using feedback from current value changes, which change inversely with a load value. Currently, in order to support the constant voltage level with variable load resistance, the so-called "tap-changing-under-load" (TCUL) transformers are used. But this transformer also does not completely solve the problem of the load resistance change with the attempt to keep the necessary phases. In the case of the single-wire system, one must increase L and decrease C proportionally to the decreasing current, as shown in figure 5.4.

The values of capacitors and inductors can be selected in accordance with the value of voltage change steps in the transformer. As a source of feedback signal, any non-contact sensor can be used.

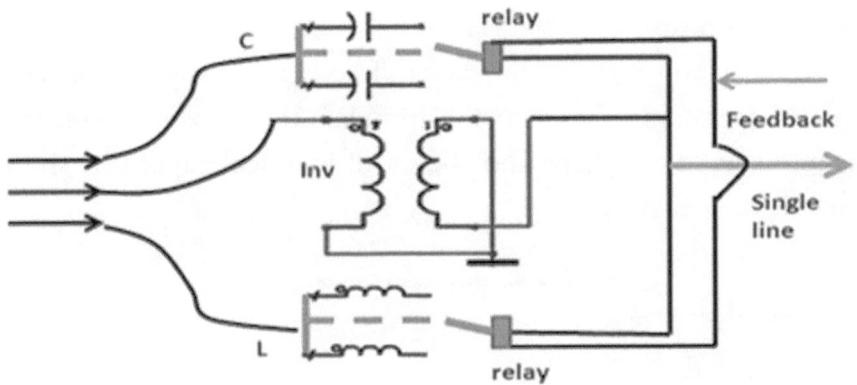

Fig. 5.4. Converter three to one

The system of the phase adjustment can be automatic.

In this case, the system consists of two parts. The first part is the control part. The feedback signal can be obtained by the known contactless method, as shown in figure 5.5.

Fig. 5.5. Contactless removal method of measuring the amount of current values

The second part includes relays, variable inductors, and capacitors. This part will be different in all systems. The system developer will prepare the table with recommendation for these elements.

The practical implementation of this circuit can be not difficult (*see* **5.6a** and **5.6b**).

The inductance set is one choke. The capacitors set are a set of identical capacitors. There are no gaps when the current is switching. This inductance with taps and the set of identical capacitors should not be very expensive.

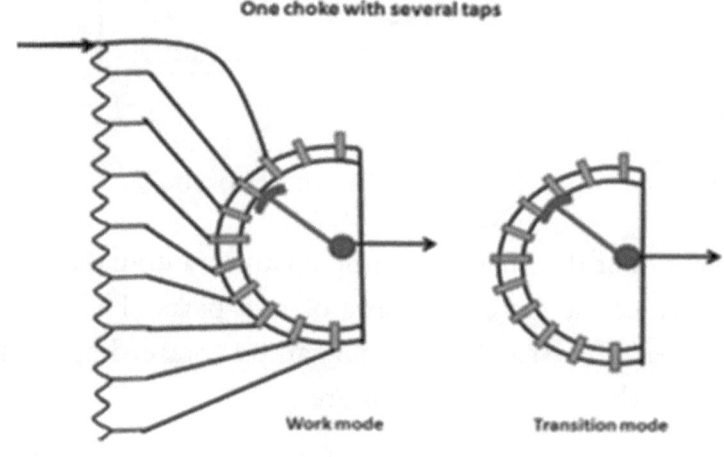

a

A set of identical capacitors

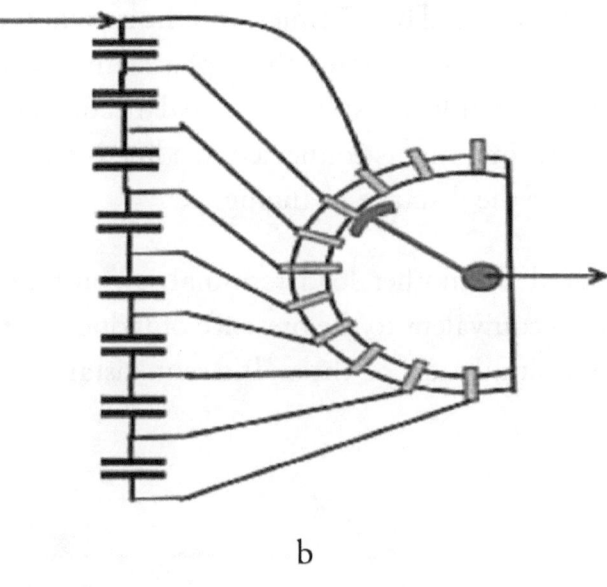

b

Fig. 5.6

As figures 5.6a and 5.6b show, in these circuits, there is no current interrupt. There is only current increase or decrease.

Thus we have shown that single-line electricity allows for transmission from a source that has a three-phase signal to a receiver with a three-phase signal by one wire instead of three or four wires. There are two important advantages as well:

- First, let's say there are three phases with wire resistances 1 ohm, current in each wire 1 A, and phase voltage 1 V. The power created in wires (power loss) will be PL = 3* 1^2 * 1 = 3 Wt. The same result can be obtained by voltage 3* (1^2 /1) = 3 Wt. But linear voltage in this system is 1.7 V. Let's make a single-line system with voltage 1.7 V also. To get the same PL, we must have resistance 3W / $(1.7)^2$ = 1.0 Ohm. So instead of three wires of 1 ohm, we will use one wire of 1 ohm. To what do we owe this large gain that has been achieved? A three-phase system alone is a poor transformer, in which the voltage is increased by 1.7 times, and the current is not lowered.

- Second, the normal long three-phase line creates reactive power. As a result this system is unbalanced. The potential of a load terminal that is connected to a line is being changed in the case of the distance changing.

The potential of another load terminal is constant and equals zero. And this is equivalent to the presence of inductance in the load. The circuit for simulation of three different distances is shown in figure 5.7.

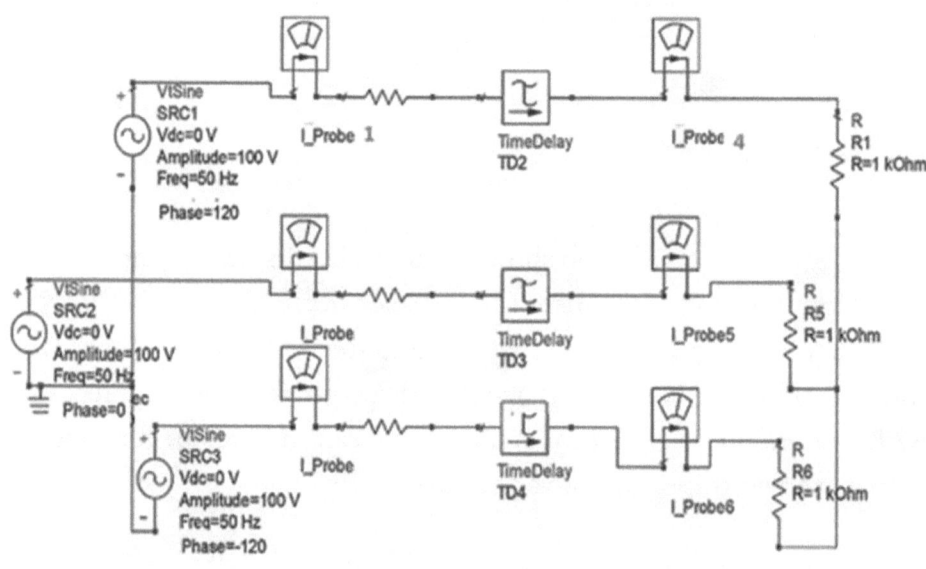

Fig. 5.7

The results of simulation with three different delays are shown in figure 5.8.

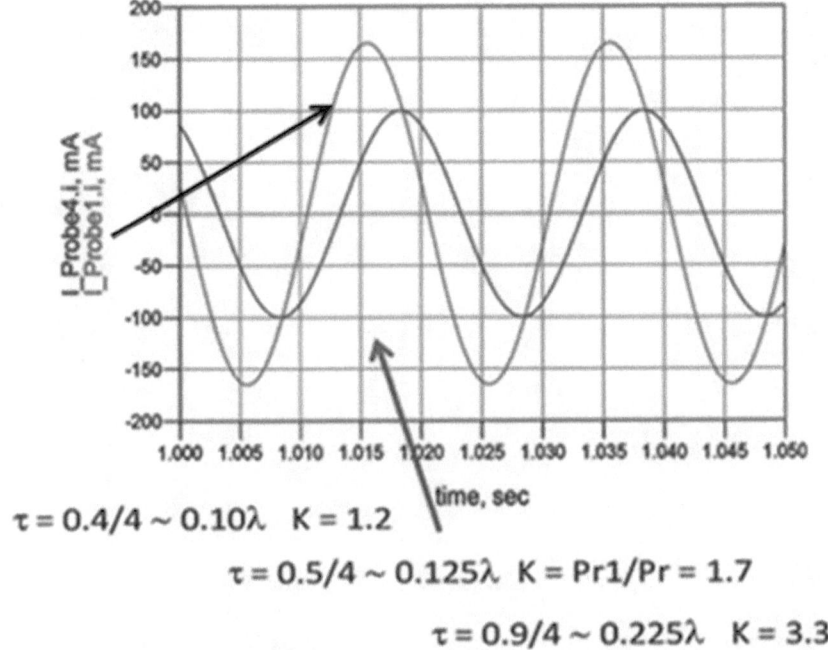

$$\tau = 0.4/4 \sim 0.10\lambda \quad K = 1.2$$

$$\tau = 0.5/4 \sim 0.125\lambda \quad K = Pr1/Pr = 1.7$$

$$\tau = 0.9/4 \sim 0.225\lambda \quad K = 3.3$$

Fig. 5.8. Three-phase system with different
delays (different lines length)

We can see that, as distance increases (delay increases), reactive power increases. Here K is the ratio between current in the generator and that in the load.

CHAPTER 6

EXPERIMENTAL SYSTEMS

The 220-V single line

In addition to conducting simulations and experiments, experimental lines were built in the interest of this research. The first working single line was made between two campuses in Lev Academic Centre (JCT). The transmitting part (source) was located in the first building. The first receiver (a 60-watt light bulb) was located in another building (*see* **6.1**).

Fig. 6.1. One wire in JCT

The second receiver (the same bulb) was located in the first building but in a different room. These buildings have different grounding points. The wires between the source and receivers are 300 metres long, and 220 V voltages are supplied to the source. The source has one single-wire output for each receiver.

The middle points of the inverters in both lines have opposite polarities. Therefore, their connection provides the necessary potential zeroing. Thus, the source and its inverters are not connected to the earth (*see* **6.2**, circuit).

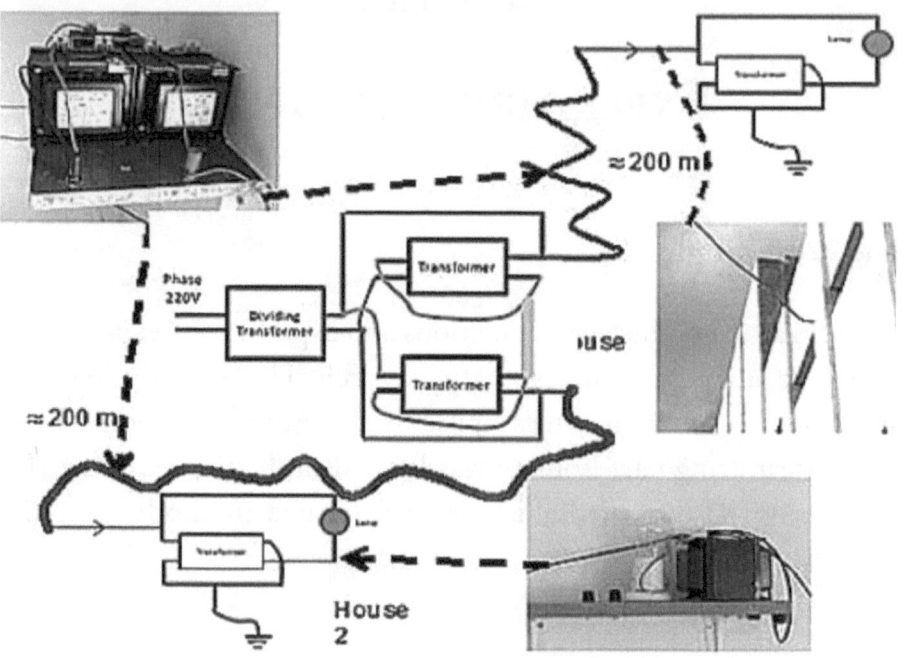

Fig. 6.2. JCT one-wire system circuit

That is, there is a grounding of both lines in the receiving ends only. One of the buildings has a separate ground for the B-line.

These lines were twice checked by a specialist sent by the Israeli Department of Energy, which has confirmed that the system operates as a single-wire line and nets no additional losses.

To demonstrate that the ground is not involved in the transmission of the electrical signal, another experiment was carried out at a frequency of 300 kHz (*see* **6.3**). In this case, it is possible to make the inverter a 500-metre long line (half period delay line) without any connection to the ground.

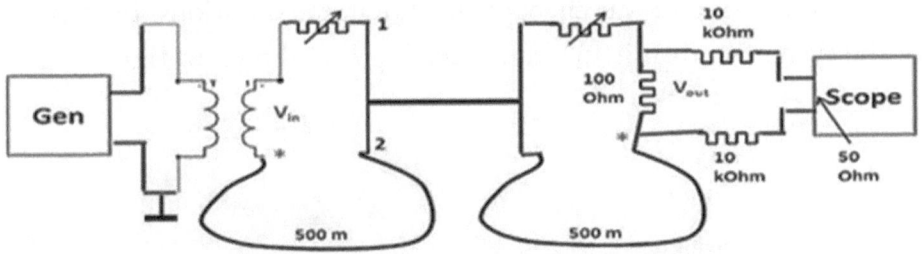

Fig 6.3. Experimental system without grounding

The 6-kV single line

Responses to the demonstration of the first experimental line were a mixture of surprise and scepticism. "It's not possible," some said. "You used ideal elements in the simulations," others pointed out. "In a real system, it won't work." Other remarks included, "There is grounding for zeroing, so the ground is another line,", "The 220 V is not a practical system," and so on.

Thus, one more experimental line was built in the centre of Israel (in the township of Tal Shahar). This line works at a voltage of 6 kV.

At first, a simulation of the proposed circuit was made (*see* **6.4**).

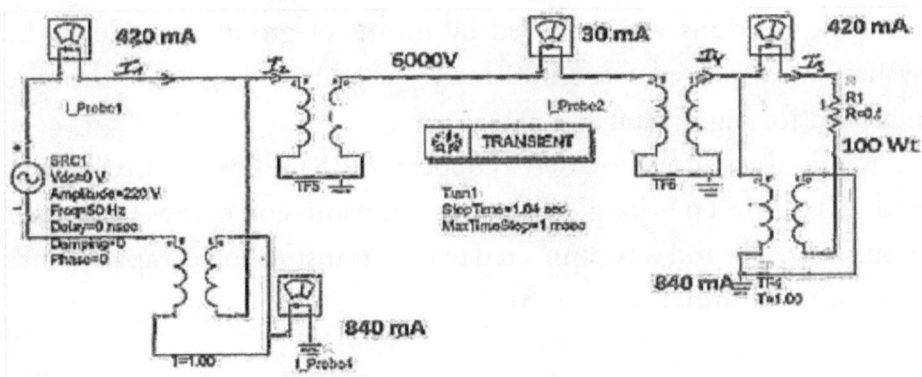

Fig. 6.4. The circuit and the simulation of
the experimental 6-kV system

Next, the system was checked in laboratory conditions. Both parts of the system (transmitting and receiving) are shown in figure 6.5.

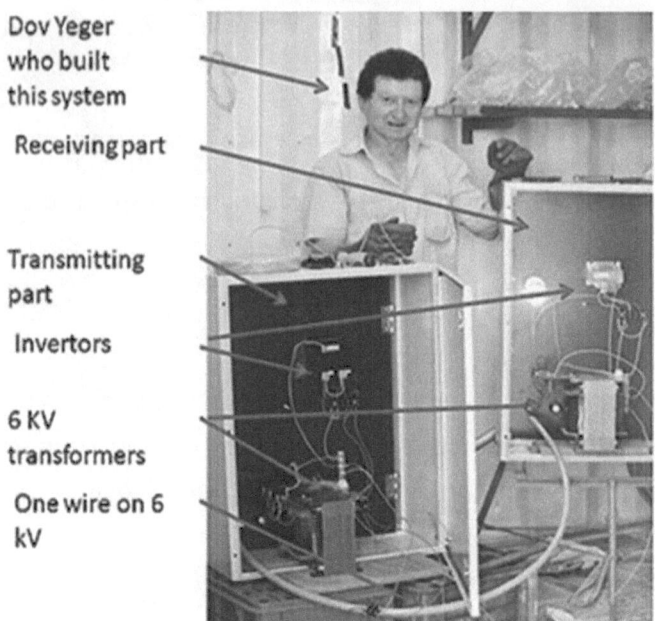

Dov Yeger
who built
this system

Receiving part

Transmitting
part

Invertors

6 KV
transformers

One wire on 6
kV

Fig. 6.5. The transmitting and receiving
portions of the system in the laboratory

During the first stage of tests, the zeroing in the transmitting and receiving portions was achieved by means of grounding. Next, the system was deployed over a 200-metre length, making use of the wire intended for high voltage.

We used a hammer to insert a copper rod, 18 millimetres in diameter and 150 centimetres long, close to the transmission compartment for grounding. The rod was connected to the \ transmission compartment's grounding terminal of (*see* 6.6).

Fig. 6.6. Normal grounding

One of the big advantages of the SLE system is the possibility that in can be used underground and in underwater lines in small cross-section pipes with no need of tunnels. Therefore, about 20 metres of the line is an underground line. This segment is inserted into a 1.5-inch plastic pipe to enable later measurement of electromagnetic fields.

The poles (metallic pipes) are strengthened with ropes to the cactus plants to prevent failure by wind blow. The first high voltage SLE moves to the receiving compartment (*see* **6.7**).

Fig. 6.7. 6-kV one-wire system in Tal Shahar, Israel

The measurement results in figure 6.8 correspond to the results taken in the laboratory (*see* **6.4**).

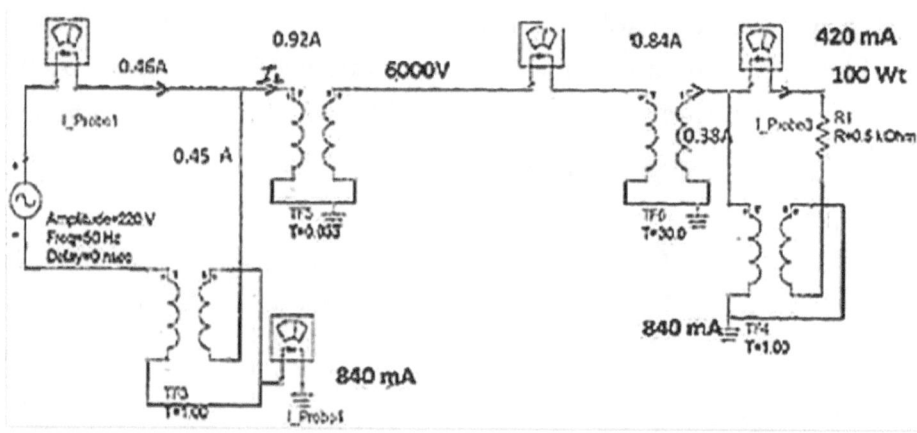

Fig. 6.8. Magnitudes of the currents in an air single-wire line

Next, we decided to show that there is no energy transmission between two grounded points. To this purpose, we switched off the grounding from the transmitting portion and switched on the nullifier, which is shown in figure 5.6. The magnitudes of all the currents have not changed.

One more experiment was carried out in the receiving part of the system. Into it we have added the converter, which is shown in figure 4.2. As a result, the three-phase signal has been obtained. The amplitudes and phases of all three currents corresponded to the necessary magnitudes.

These experiments are clear – an SLE system, with a 6kV-voltage on a 50-Hz frequency is working in Tal Shahar, Israel, and results in no additional losses. This system can work without real grounding but with, instead, a nullifier (*see* **3.10**). Therefore, a ground is not the second line. The load and source of a single-line system can be a three-phase system. The system in Tal Shahar, Israel, was checked by several groups of specialists from Israel and other countries.

During the course of these tests, it was shown that the transition from real grounding to a nullifier application does not cause any changes in the 6-kV single-wire system.

CHAPTER 7

SINGLE-LINE GREEN ENERGY SYSTEM

Common two-wire structure

Nowadays, the majority of "green" (environmentally safe and renewable) energy systems are being built as systems of direct current, and all the sources are being connected with an energy storage device by means of two-wire lines. Systems of wind energy are being built in the same way, though every source (wind generator) generates alternating current (*see* **7.1**). The main reason for using the direct (but not alternating) current is the synchronization problem of all the sources. Therefore at output of every source, a rectifier is used.

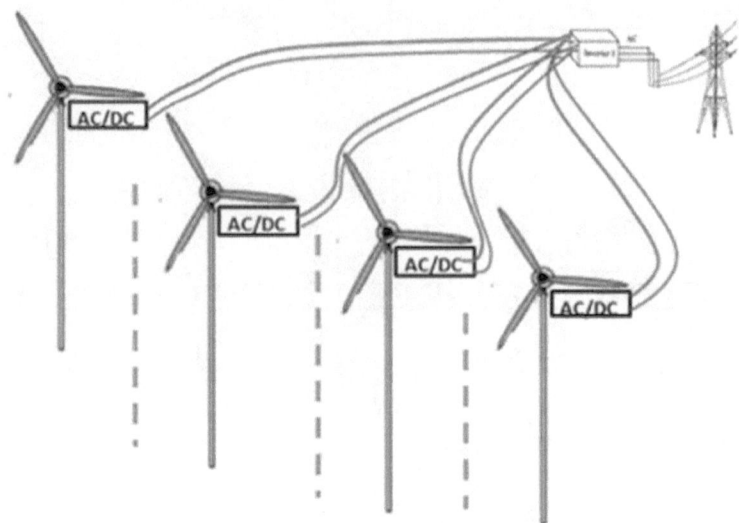

Fig. 7.1. Common two-wire structure

What follows is a description of how to apply the single-wire method of energy transfer to the various "green" sources of energy. As shown in the previous chapters, convertors from two to one and three to one, as well as the nullifiers, are easier to perform on alternating current. And they are still simpler, if the frequency of the current is greater than 10 kHz. As is shown below, in this case, the converter can be performed as a bifilar line. But the synchronization problem remains.

For a solution to this problem, we decided to use not direct current but quasi-direct current.

The system proposed here does not include rectifiers from AC to DC and inverters from DC to AC, batteries, groundings, and internal synchronization blocks. The system consists of blocks as sources and blocks for collecting. Each block source includes the source and generator of frequencies, for example100 kHz. The source, which contains a generator, is built in accordance with the single-wire method described in chapter 2, using a delay line on the half wave.

Let's name such a source one-pole source (OPS). On these frequencies, the delay line can be made as a bifilar coil.

The transition from a direct current source to the sources of quasi-direct current is shown in figure 7.2

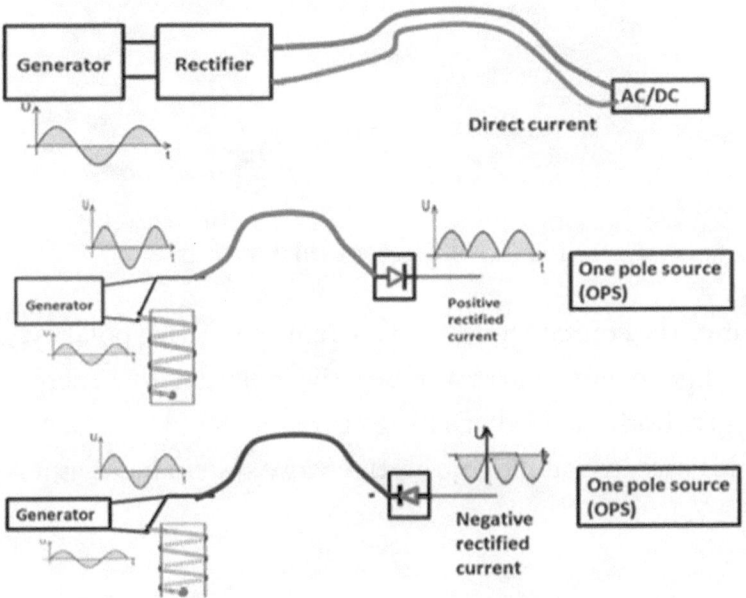

Fig 7.2. Direct and quasi-direct currents

The converter or the delay line can be performed in the form of the known bifilar line (*see* **7.3**).

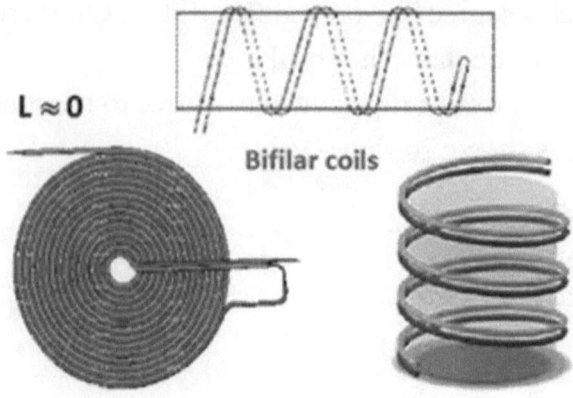

Fig 7.3. Different bifilar coils

Half of the sources generate a current of positive polarity, and the other half generates a current of negative polarity. The energy storage system gets both quasi-alternating currents.

In this case, the single-line green energy system looks as it is shown in figure 7.4.

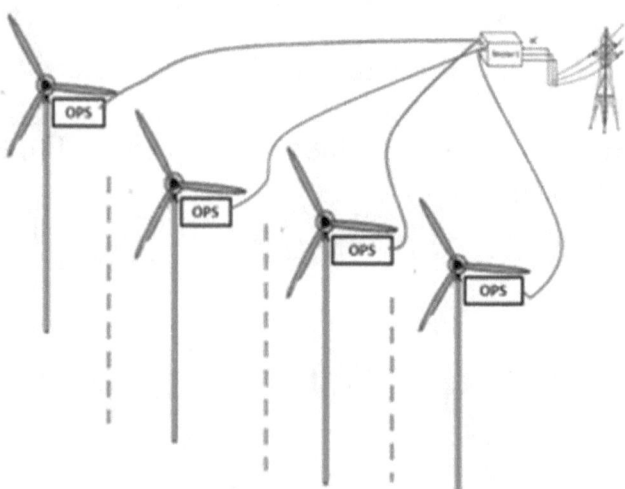

Fig. 7.4. Proposed one-wire structure

Simulations

As a basis for the rectifier in collecting block, we can use a normal balanced single-phase full-wave rectifier (*see* **7.5**).

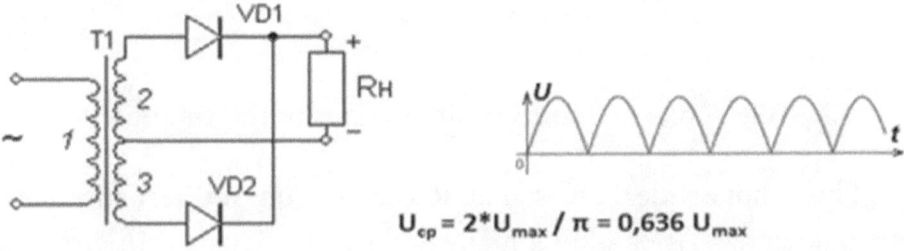

$$U_{cp} = 2*U_{max} / \pi = 0{,}636 \, U_{max}$$

Fig. 7.5. Rectifier and its output current

The simulation of a system with one block source was made by the circuit shown in figure 7.6.

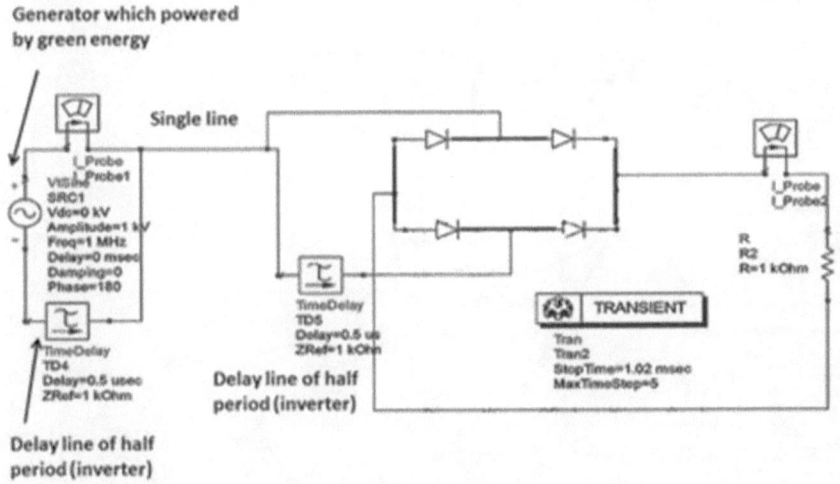

Fig. 7.5.1. Proposed circuit for simulation

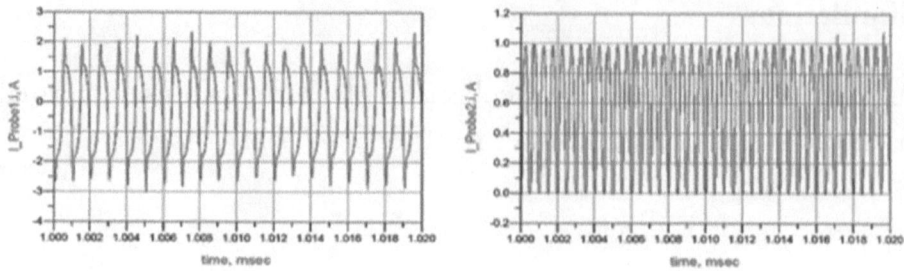

Fig. 7.6. The results of simulations of the circuit

This is not an ideal DC signal. Rather, it is quasi-direct current or one-pole signal. After adding many of the same signals with different delay, we can get the correct DC signal. These delays are created due to the different distances between block sources and the collecting block.

The circuit in figure 7.7 shows the simulation of a system with five block sources.

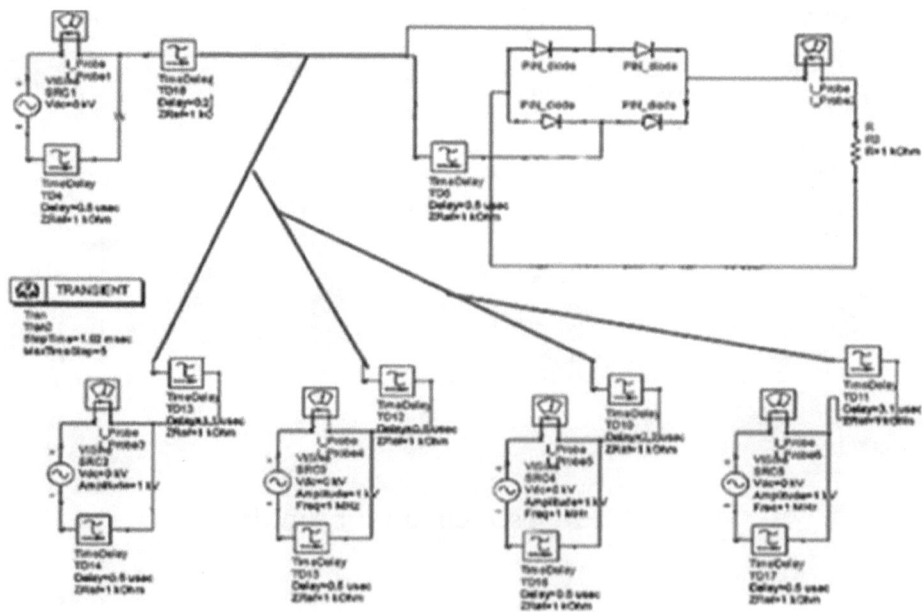

Fig 7.7. Circuit for five sources

Each signal of the source blocks has a different delay that corresponds to distances from 0.1 kilometres to 1 kilometre. The results of the simulations – a quasi-constant signal – are shown in figure 7.8. It is obvious that, by connecting low-frequency filters to both inputs of the storage device (a battery or a generator), we get the normal two-wire signal of direct current.

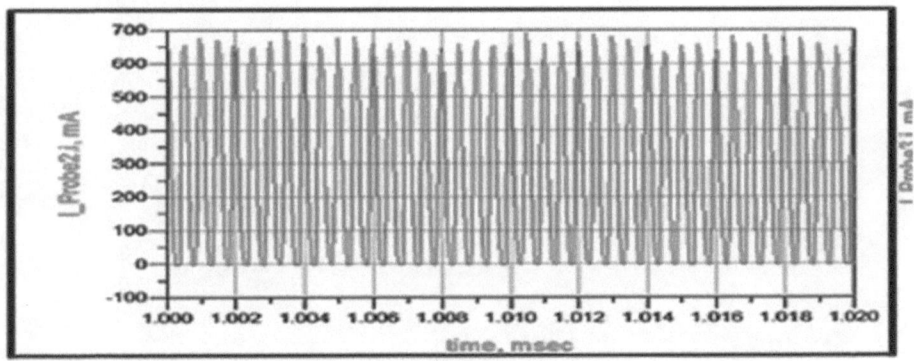

Fig 7.8. The current of five sources

Notably, we used two times fewer wires, created a simpler system of conversion of the energy sources, and excluded grounding problems in direct current systems.

CHAPTER 8

INTERFERENCE AND LOSS

Electric field and corona effect

High voltage between wires creates an electric field and corona effect. In addition to the possible harmful effects on the environment, these factors lead to additional losses. Remember that a linear voltage in a three-phase system has a square root of three times more than the voltage in each phase.

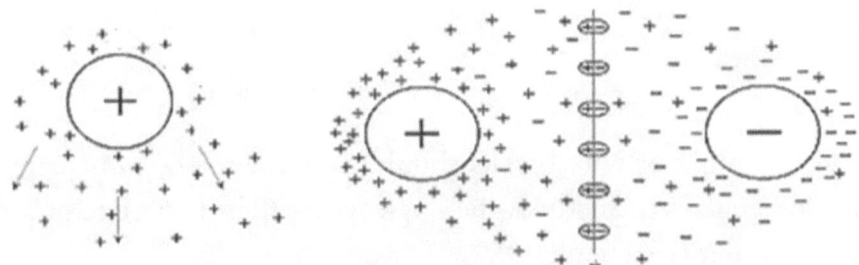

Fig. 8.1. Corona effect

As a result of high voltage, the ionization rate increases, and, consequently, the current crown and loss of energy are increasing. This regime is called bipolar corona (*see* **8.1**, right side).

Appearance of corona creates not only additional losses in wires, but also additional distortions of the initial sinusoidal form of the current. Neither of these is acceptable for the alternating current networks that are now in use.

This problem almost does not exist in the B-Line system (*see* **8.1**, left side).

A few possible options for short circuits exist. Wire breaks and short circuits are serious problems in many regions of the world. Nowadays the servicing of a two-wire system is rather expensive. I am certain that the future systems of electric power transmission will be single-wire systems. Figure 8.2 demonstrates that transitioning to a single-wire method will sharply reduce the number of short circuits. Fewer wires equals fewer short circuits equals fewer accidents.

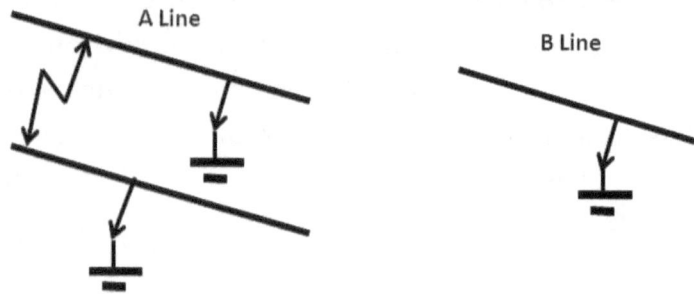

Fig. 8.2. Short circuit cases

The number of wire breaks should decrease to the same extent. With the single-wire method, the single wire will be laid underground or under water (see chapter 9).

Power losses

The main advantage of single-wire systems is the use of one wire instead of two or four, which leads to a drastic reduction in the cost of the electrical system (Bank et al. 2012).

Another important advantage is the decrease of losses in the transmission of electrical energy. Calculations and simulations show that the mutual influence of the two relatively closely spaced conductors with currents of opposite polarity results in an increase in resistance of both wires (*see* **8.3**).

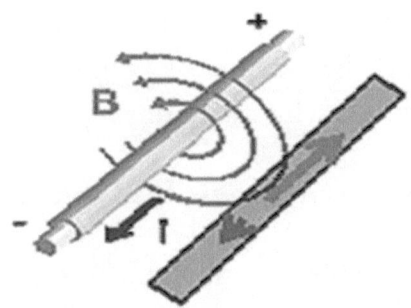

Fig. 8.3. The mutual influence of two wires

This problem is studied in detail with regard to cables called *twisted pair*.

Twisted pair cabling is a type of wiring in which two conductors of a single circuit are twisted together for the purposes of cancelling out electromagnetic interference (EMI) from external sources. For instance, this cabling cancels electromagnetic radiation from unshielded twisted pair (UTP) cables and crosstalk between neighbouring pairs. Attenuation of crosstalk is achieved on a Cat 5 (category 5 cable) channel. (A channel may consist of up to 90 metres of horizontal cable, one or two transition connectors on each end of the horizontal cable, and up to 10 metres of user patch cables for a total maximum length of 100 metres) (Cabling Installation & Maintenance Magazine 2014).

In the three-phase systems of electrical energy transfer, the method of phase splitting is being applied. This means that energy is being transmitted by three separate wires. This not only provides a decrease of losses due to corona and more convenience at installation and mounting of the line, but it also has one more important advantage related to a decrease of the line inductance. Owing to this decrease, the transmission capacity is increased. That is especially important for lines of extra-high voltages, which are purposed for transfer of very large amounts of power. With increase of the distance between wires, the inductance of the line per unit of length simultaneously decreases. Therefore, a small departure from the optimal distance results in a very small increase of the maximal field intensity. This problem does not exist with the single-wire method.

The influence of electromagnetic fields on human beings

There is a serious problem with the assessment of the impact of radiation from high-voltage electric lines and transformers on human health. There are norms, recommendations, and measurement methods for parameters like electric field strength, magnetic field strength, and magnetic induction. However, these parameters do not allow for the estimation of the power level that enters the human body.

The opponents of a single-wire method often say that, in the case of the usual two-wire or three-wire methods, the electromagnetic field being created by several wires is mutually compensated. But if there will be only one wire, the influence of the field it creates will be greater. However, these dangers are probably exaggerated.

Consider that the electric wire is a transmitting electric aerial, and the human body is the receiving antenna. From the theory of aerials, it is well known that efficiency of the aerial is very low if the length of the aerial is much less than a quarter of the wavelength. Here we have just that very case. The wire above the head of a human being and all the more the height of human body is many times less than quarter of wavelength. Note that the wavelength on a frequency of 50 Hz is equal to 6.000 km.

Powerful transformers that wind using a great number of turns, as well cores with high magnetic permeability can create a strong electromagnetic field. Such transformers can represent a magnetic antenna with a large effective height.

Figure 8.4 shows electric energy entering an apartment house in the centre of Jerusalem, using a step-down transformer.

Fig. 8.4. Electric energy from a step-down transformer entering a Jerusalem apartment

Clearly, those transformers will be used in many-wire systems, as well as in single-wire systems. But the single-wire method allows for the supplying of electric energy via underground routes. In this case, the transformers or switching devices can be located at ground level, or on a lower level than in the setup shown in figure 8.4 – in other words, not in front of the windows of an apartment house.

CHAPTER 9

UNDERGROUND AND UNDERWATER SINGLE-LINE SYSTEMS

Two problems of an underground three-phase system

First, the three-phase system requires a certain distance between its wires. Decreasing this distance causes large additional losses. In order to decrease these losses, one needs to build very expensive tunnels (*see* **9.1**).

Left: 345 kV XLPE project – Cement vault visible with two chimneys extending up to be level with the future road surface.
Right: 138 kV XLPE project – Bottom half of pre-constructed vault positioned in trench.

Fig. 9.1. Tunnels for the underground three-phase system

The second problem is cable capacitance per unit length (Xue et al. 2012).

The capacitance of each wire increases in the presence of other wires. This increased capacitance causes a large reactive power. In order to prevent this in extended systems, intermediate stations are used, which provide reactive power compensation.

Another solution is to transmit electrical energy by DC. However, while transmitting via DC can solve these problems, it is a very expensive solution.

The reactive power in electrical systems is a well-known problem. Our next simulation (*see* **9.2**) will demonstrate the problem. We'll take a capacitor and connect it parallel to the load.

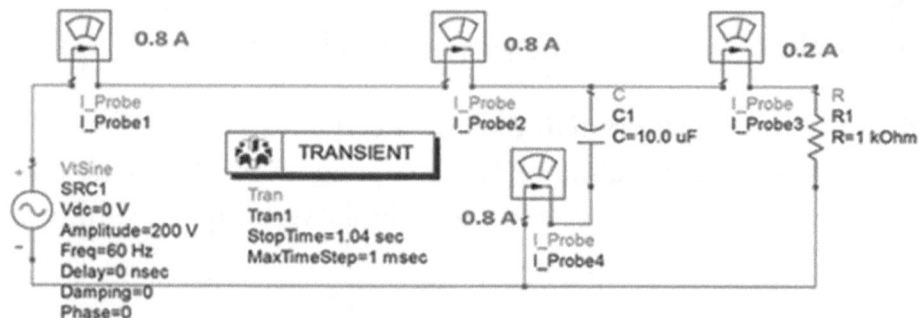

Fig. 9.2. Simulation of a three-phase system with capacitance being connected parallel to the load

A huge reactive power results (the input current increases from 200 mA to 800 mA). The same currents are obtained if the second wire is connected to ground.

On the other hand, in a single-wire system, the same capacity creates a much smaller reactive power (*see* **9.3**).

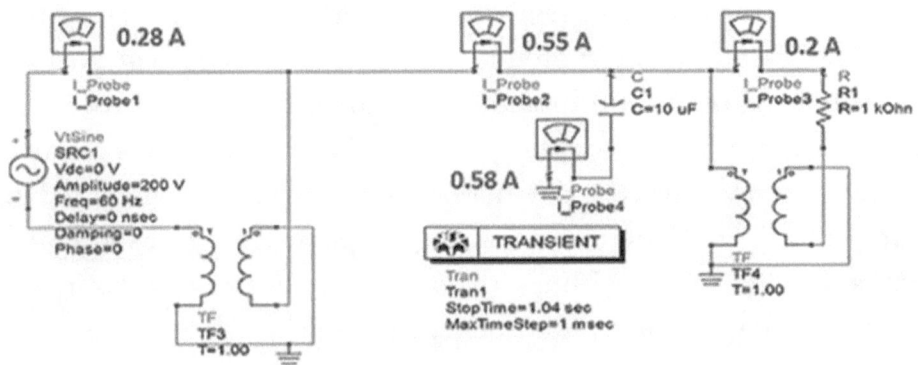

Fig. 9.3. Simulation of a single-wire system with capacitance being connected parallel to the load

The largest influence has the linear or charge capacitance. To a large degree, it is the capacitance between either cable or between the cable core and the cable braiding (*see* **9.4**) (Bank and Haridim 2009).

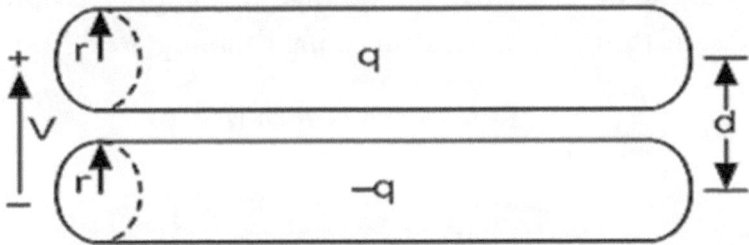

Fig. 9.4. The second wire's influence on the cable capacitance

Usually it is implied that the cable transmits a one-phase signal. One pole of the phase is connected to the cable's central core, and the second one is connected to the cable's braiding. The cable's braiding can be grounded at input, as well as at output. However, between the input and the output, the wire is not grounded.

In the single-wire system, there is no (or almost no) linear capacitance. Therefore, it is possible that there is no need to compensate anything. That is, in this respect, the single-wire system does not differ from the DC system.

In the single-wire system, there is nothing to connect to the braiding; even if it will be grounded, it is still not a two-wire system.

Following our useful discussions, I would like to formulate and discuss reactance compensation problems. They are a result of reactive power. In general, reactive power appears if a reactance is connected in parallel to the load. Serial reactance gives phase-shift and additional losses. In high-voltage airlines, there is a large degree of capacitance between wires – in other words, parallel to the load. Therefore, there are compensation devices in intermediate stations.

One of the important advantages of an SLE is the absence of capacitance between wires and the very small capacitance between wires and ground. Therefore, it is possible that there would be no need for intermediate stations. In underground and underwater lines, there is capacitance between wire and ground. So we need to use intermediate compensation devices. I don't see this as a large problem. It is possible to develop small devices, like an amplifier in optical line system (something similar to connecting coupling; *see* **9.5**).

Reactance compensator

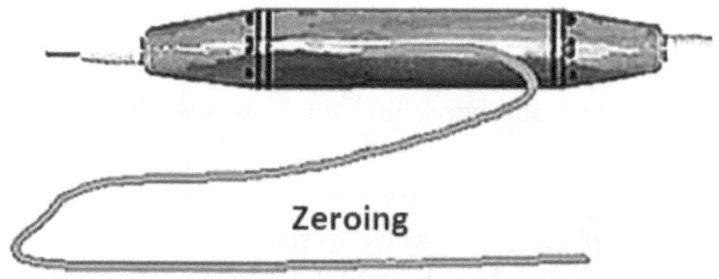

Zeroing

Fig. 9.5. Possible underwater zeroing

One must add a wire nullifier. In an underwater system, this compensator will be in saltwater on the bottom of the seabed.

At the request of one electric company, we made the simulation of a single-wire system for the transmission of electrical energy (1.3

MW) in line, which consists of three sections (300, 400, and 500 kilometres).

The simulation demonstrated that, by using a serial inductance and parallel capacitor, we can eliminate the reactive power (*see* **9.6a** and **9.6b**).

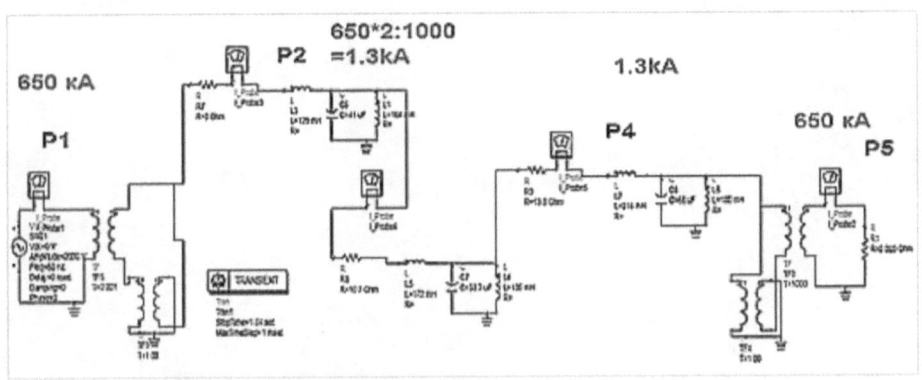

a

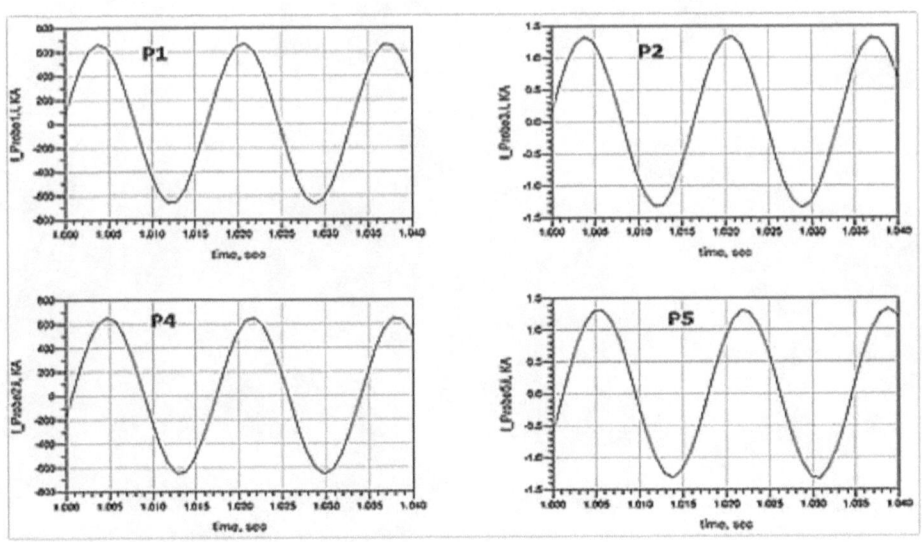

b

Fig. 9.6. Simulation of a three-section, single-wire underwater line

Thus, at application of a single-wire method for the overhead (air) systems, the intermediate stations for reactive power compensation will likely not be necessary. In the case of underground or underwater systems, the small compensating elements can solve the problem conditioned by the influence of running inductance and linear (charge) capacity.

CHAPTER 10

HIGH-FREQUENCY SINGLE-LINE SYSTEMS AND ANTENNAS

Line on high frequency

We will now illustrate that the idea of the B-Line is also correct for high frequency. First, we compare a normal long line with a characteristic impedance of 300 ohms with a B-Line on a frequency of 1.1 GHz.

On high frequency, it is possible to produce an inverter as a delay line, where its length equals half the wavelength (Bank and Haridim 2009) or a one-port strip line (*see* **10.1**).

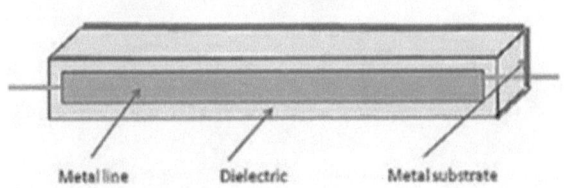

Fig. 10.1. One-port strip line structure

A simulation of one long wire line was conducted using this strip line, which is normally equivalent to a 300-ohm long line. The circuit A-Line (mode 1) and B-Line (mode 2) are shown in the graphs in figures 10.2 and 10.3. as a load is implemented to the source with a resistance of 300 ohms and no voltage. In the same figures, one can see parameters S11 and S21.

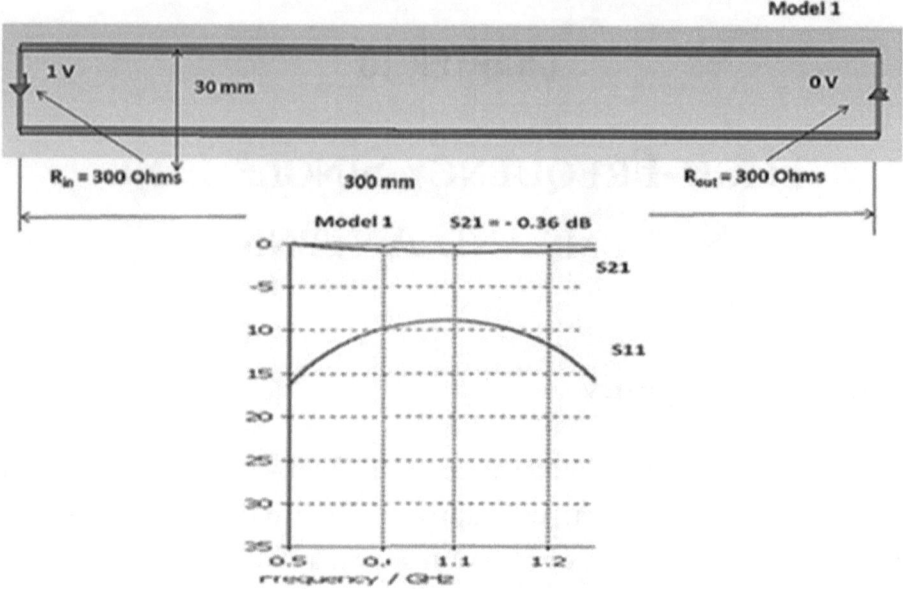

Fig. 10.2. Normal long line and its S parameters

Model 2

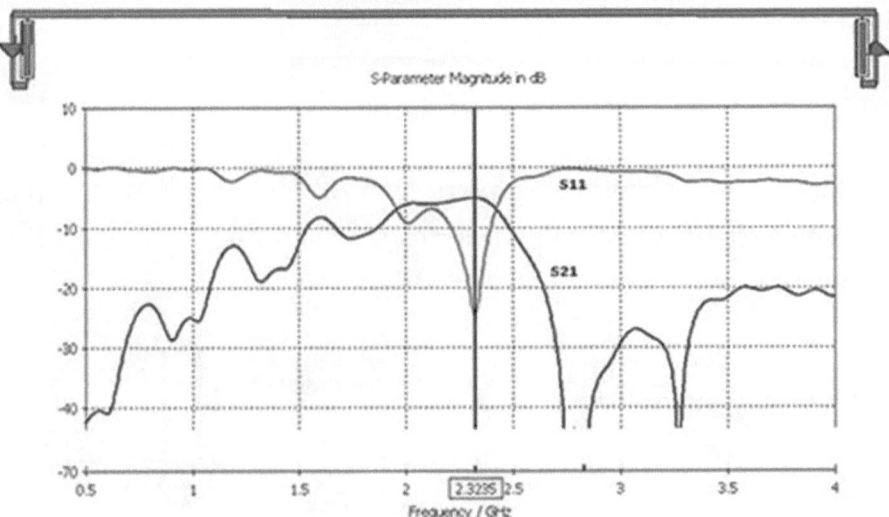

Fig. 10.3. Proposed one line and its S parameters

The matching long line has infinite bandwidth. This has an advantage but also a disadvantage. The advantage is that you can pass through a long line of multiple signals with different frequencies. However, in a real system, there is always some noise. Even if the noise is weak, in an infinitely wide band, the noise will still be infinitely large (this is true, of course, only if the noise is white). Although you can, of course, apply a filter at the input of the receiver, doing so is often problematic. The filter introduces loss and increases the noise factor.

The proposed single-wire system (B-Line) is a selective system. The disadvantage of the B-Line is a need to change the delay line in case of change of frequency. The B-Line is compatible with the source and load, and in this sense, there is no difference between it and the usual long line. It is selective but rather broadband. It has no requirements in terms of symmetry, which is often a problem in the prior-art systems when the use of long lines inside the apparatus can result in different influences on each wire. Maybe today's systems are better?

Using B-Line on antenna construction

B-Line principle allows the construction of a monopole with dipole parameters (MB antenna) (Bank et al. 2012). Figure 10.4 clarifies this idea.

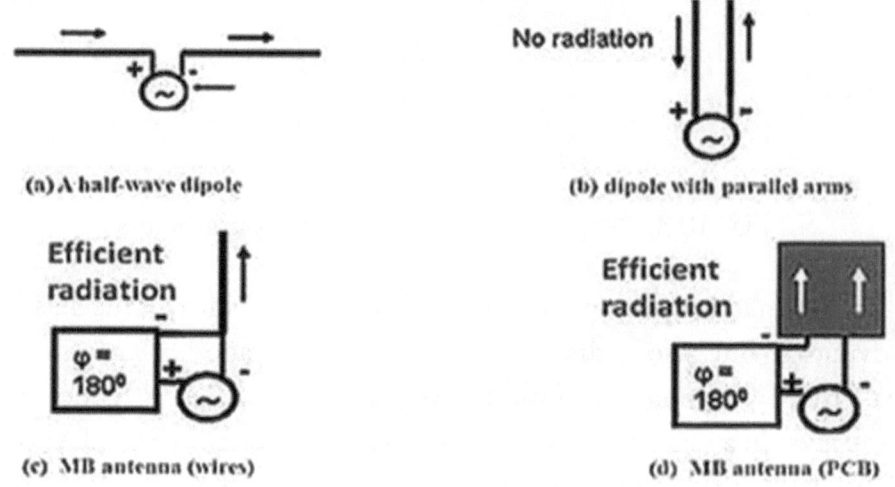

Fig. 10.4. Single-line idea and MB antenna

The MB antenna is described in detail in an article called "Highly Effective Handset Antenna" published in the *International Journal of Communications* (Bank et al. 2012). It is shown that, in comparison to a dipole, the MB exhibits higher gain and efficiency. The enhanced performance of the MB is attributed to the fact that the MB is based on a shorted two-wire line approach, which is in contrast to the conventional linear antennas with open-ended arms.

One important advantage of the MB antenna lies in the fact that its promising characteristics are achieved while eliminating the need for a separate antenna to be implanted in a mobile unit.

Another advantage of MB is its circular radiation pattern.

Improving the MB gain by 3 dB in the transmitting mode and by 3 dB in receiving mode allows doubling the communication range.

The conventional dipole is a balanced antenna, so that its arms are connected to the leads of a balanced source. In this case, the currents on the dipole arms are in anti-phase, and since the arms are in opposite directions, the emissions from the arms add constructively.

In the case of the MB antenna, the leads of the balanced source are both fed into a single radiator. The currents from the leads are made in-phase by means of an inverter (a 180-degree phase shifter) inserted in the path of one of the source leads. In this manner, the current in the (single) radiating element of the MB antenna is doubled, compared to a dipole antenna.

The printed circuit board (PCB) (**this abbreviation is used wide in cell phone techniques**) mobile handsets can be used as the radiating element of the MB antenna. In this case, one has merely to add a 180-degree phase shifter to one lead of the signal source. The MB antenna actually removes the need for a separate antenna such as a planar inverted-F antenna (PIFA) to be installed on the PCB, and consequently, it eliminates any potential adverse effects of the PCB on antenna performance.

The MB antenna can be designed for multiband operation and can provide very broad bandwidths. In addition, the idea behind the MB antenna can be applied to transparent antennas of large areas. These properties make the MB antenna a promising antenna for energy harvesting applications.

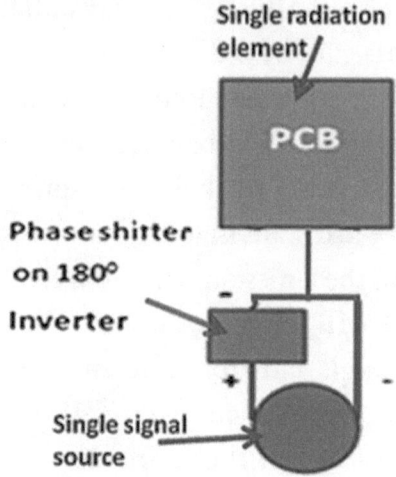

Fig. 10.5. Connection of source and radiator in MB antenna

In some cases, using PIFA antennae in small transducers is problematic (*see* **10.6**).

PIFA with PCB

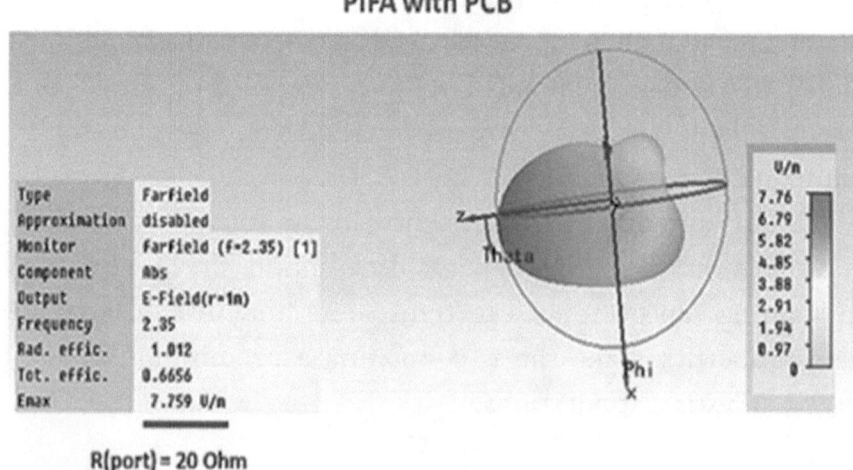

R(port) = 20 Ohm

Fig. 10.6. Far-field distribution of PIFA

The use of an MB antenna enables the building of a transducer without antennae, as a radiator will be PCB (*see* **10.6**).

MB Antenna with Strip Line as a converter

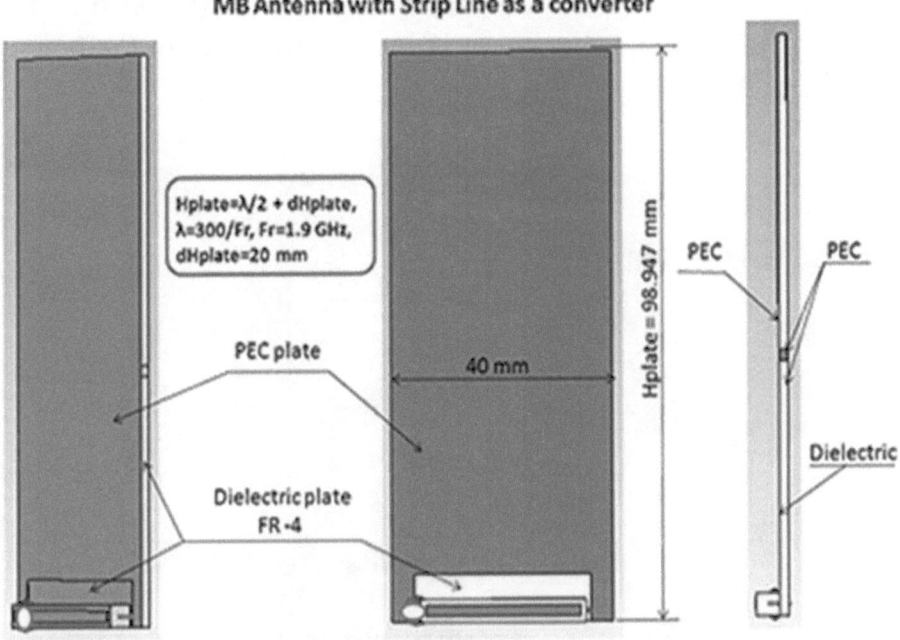

Hplate=λ/2 + dHplate,
λ=300/Fr, Fr=1.9 GHz,
dHplate=20 mm

PEC plate

40 mm

Hplate = 98.947 mm

PEC

PEC

Dielectric plate
FR-4

Dielectric

Fig. 10.6. MB antenna in a cell phone

Figure 10.7 gives an example of a MB antenna and its parameters.

MB antenna

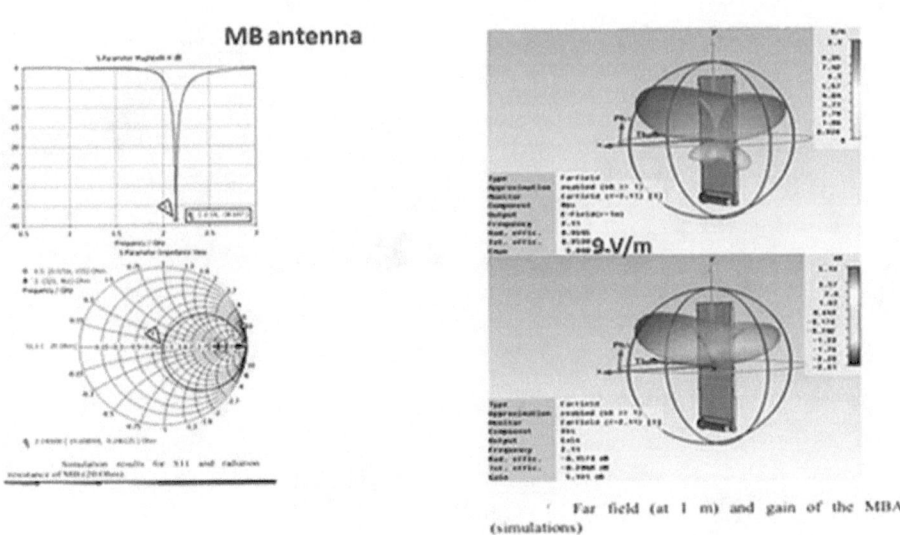

9.V/m

Far field (at 1 m) and gain of the MBA (simulations)

Fig. 10.7. Simulation results of MB antenna on figure 10.6

The MB antenna principle allows for the development of a telephone in a watch as shown in figure 10.8 (but not in the pocket via Bluetooth).

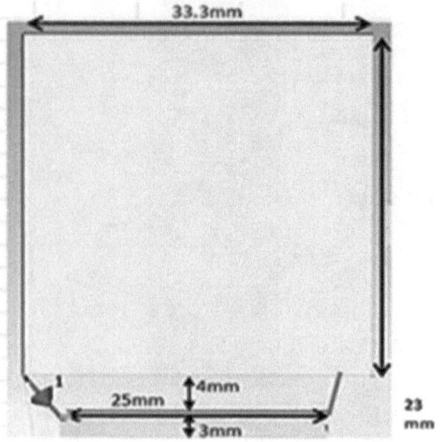

Fig. 10.8. Very small MB antenna

The results of this model simulation are shown in figures 10.9 and 10.10.

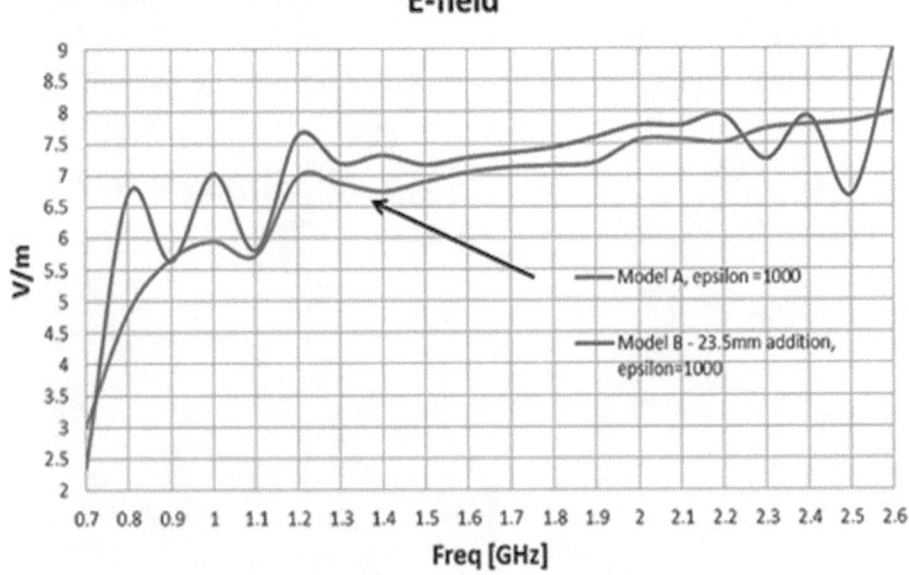

Fig. 10.9. E-field simulation of MB antenna on figure 10.8

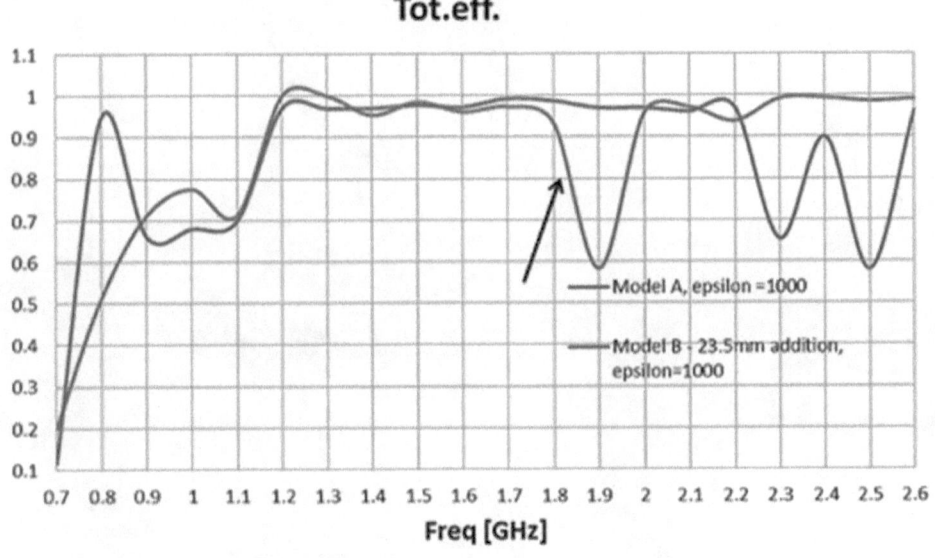

Fig. 10.10. Efficiency simulation of MB antenna on figure 10.8

Thus, the MB antenna is a new type of antenna. It is neither dipole nor monopole. It consists of one metallic surface as the radiating element and a phase shifter. In cellular handsets, the existing PCB can be used as the radiating element so that only a phase shifter is needed. The characteristics of the MB antenna are similar to those of a dipole and even better.

Conclusion

The single-line electric (SLE) system has been proposed, and, after simulations, its capability has been verified. Now in Israel, two SLE systems – one operating with 220 volts and the other with 6 kilovolts – are functioning. SLE can operate at all frequencies, including 50 and 60 hertz, as well as DC; with all levels of power; and with all voltages. Electrical energy in an SLE AC system is not transmitted through the ground.

The SLE (B-Line) method enables:

- The realization of any electrical system, including smart systems, as a single-line system
- The connection of a three-phase generator and a three-phase motor with one wire only
- An SLE AC system that doesn't require inputting electrical current to the ground surface
- The transmission of electrical energy from the place of generation to the place of implementation (and, thus a cheaper system, as in many cases, producing electricity at the place of gas production and transferring it to the place of consumption via single wire under the ground will be more profitable)
- A significant reduction in the number of wires on the globe
- A significant reduction in the cost of construction of high-voltage electrical systems

- The use of underground and underwater power lines for energy transmission without intermediate stations
- A reduction in the number of accidents associated with electrical systems
- A reduction in energy losses during transmission as a result of using one wire instead of the common three-phase wires
- The construction of a new type of an antenna (an MB antenna), which will allow for the creation of a receiving transducer without an antenna

A separate conclusion concerning the problem of zeroing

The proposed method of zeroing gives hope for the possibility of creating AC systems that don't require the insertion of electrical current into ground surface.

Chapter 3 demonstrates that, in future, power plants will be able to operate using single-wire systems, where zeroing may not be required at all.

Some research had proved that the danger for living creations created by current injection into the earth is strongly exaggerated.

Chapter 3 also gives an example in which the current strength in a line (and consequently the magnitude of the current that requires zeroing) is 10 ampere. It is known that the ground space within which one can detect a current is a cylinder of 10 metres in diameter and 10 metres of depth. That means that its volume is approximately 800 cubic meters. If the resistance of the grounding is even 10 ohms, a 10-ampere current will emit 1.000 watts of power. Thus, the average amount of power in a single cubic metre will be 1.25 watts. In addition, as is being supposed in chapter 3, this power is mainly spent on the creation of a very weak electromagnetic field.

For a collection of all available information on the single-line electric (SLE) method, including articles, presentations, and YouTube films, visit Michael Bank's website, located at www.ofdma-manfred. com. For technical questions, please send queries to bankmichael1@ gmail.com. For questions referring to possible collaborations, please write to larry@sleint.com.

ACKNOWLEDGEMENTS

Author is grateful to the company SLE International and Jerusalem College of Technology for the help in patenting and manufacturing of experimental systems.

Special thanks to Arkadiy Maltinsky and Victor Lander for their invaluable help in the preparation of *Quite Another Electricity*.

I extend a sincere thank you to all who expressly or by implication promoted the development of the single-wire method. The majority of these people were (and probably still are) against this idea. But they have helped too, as their input helped me see what was necessary to explain and what questions needed answers.

I am grateful for the creative contributions of Jacob Gitman, Alexander Bronshtein, Motti Haridim, Dov Yeger, Ely Zborovsky, Yury Shalit, Alexander Shubov, Ludmila Pak, Miriam Bank, David Chrnomordik, Roman Sokolovski, Yuri Okunev, Ted Rechels and, and many of my students, whose simulations results are included in *Quite Another Electricity*.

ACKNOWLEDGEMENTS

REFERENCES

Avramenko, Stanislav, and Avramenko Konstantin (1994), "Solid State Space-Energy Generator," *New Energy News* (August). (The AFEP experiment was based on the 1993 Russian patent application by Stanislav and Konstantin Avramenko, PCT/GB93/0096.)

Bank, M. (2012), "Single Wire Electrical System," *Engineering* 4: 713–722.

——— (2014) "Balanced and Unbalanced Single Wire," *International Journal of Emerging Technology and Advanced Engineering* 4 (10) (October) <www.ijetae.com> (ISSN 2250-2459, ISO 9001:2008 Certified Journal).

——— and Haridim, M. (2009), "A Printed Monopole Antenna for Cellular Handset," *International Journal of Communications* 2 (3).

———, Haridim, M., Tsingouz. V., and Ibragimov, Z. (2012), "Highly Effective Handset Antenna," *International Journal of Communications* 2 (6): 80–87.Bank, Michael (1982), Bank M. *Parameters of Household Receiving-Amplifying Equipment and Methods of Their Measurement*, Radio and Communication Publisher, Moscow, (Russian). 1982

Bettine, Frank (2002), "Proposal to Use Single-Wire Ground Return to Electrify 40 Villages in the Calista Region of Alaska," 2002 Energy Conference, University of Alaska <http://cem.uaf.edu> modified 10 Oct. 2002; accessed 10 Sept. 2008.

Cabling Installation & Maintenance Magazine (2014), April 2014.

Gerke D., and Kimmel, W. (1998), "The Mysteries of Grounding," *IEEE EMC Society Newsletter* (Summer): 13–14.

Goubau, Geog (1950), "Surface Waves and their Application to Transmission Lines," *Journal of Applied Physics* 21.

Morrison, Ralph, and Lewis, Warren H. (1990), *Grounding and Shielding in Facilities* (New York: Wiley-Interscience), 47–49.

Weber, Ernst, and Nebeker, Frederik (1994), *The Evolution of Electrical Engineering* (Piscataway, New Jersey: IEEE Press, 1994).

Whitlock, Bill (2005), *Understanding, Finding, and Eliminating Ground Loops in Audio and Video Systems* (Chatsworth, California: Jensen Transformers, Inc.) <http://www.historyofrecording.com/support-files/whitlock-generic-lecture.pdf> accessed 18 Feb. 2010.

Xue, Yiyan, Finney, Dale, and Le, Bin (2012), "Charging Current in Long Lines and High-Voltage Cables Protection," 39th Annual Western Protective Relay Conference, October.

Recent articles by Dr Michael Bank, author of *Quite Another Electricity* include:

- Bearbeitung der schallinformation im menschlichen gehorsystem und in technischen anlagen Rundfunktechnischen Mittelungen 2 (1992), 53–65.
- "On Increasing OFDM Method Frequency Efficiency Opportunity," *IEEE Transactions on Broadcasting*, 50 (2) (2004), 165–171.
- "Redundancy Versus Video and Audio Human Perception," *International Journal of Communications* 1 (4) (2007), 180–195.
- M. Bank, M. Haridim, V. Tsingouz, and Z. Ibragimov, "Highly Effective Handset Antenna," *International Journal of Communications* 6 (2) (2012), 80–87.